Springer Tracts in Modern Physics
Volume 200

Managing Editor: G. Höhler, Karlsruhe

Editors: J. Kühn, Karlsruhe
Th. Müller, Karlsruhe
A. Ruckenstein, New Jersey
F. Steiner, Ulm
J. Trümper, Garching
P. Wölfle, Karlsruhe

Available online at
SpringerLink.com

Starting with Volume 165, Springer Tracts in Modern Physics is part of the [SpringerLink] service. For all customers with standing orders for Springer Tracts in Modern Physics we offer the full text in electronic form via [SpringerLink] free of charge. Please contact your librarian who can receive a password for free access to the full articles by registration at:

www.springerlink.com

If you do not have a standing order you can nevertheless browse online through the table of contents of the volumes and the abstracts of each article and perform a full text search.

There you will also find more information about the series.

Springer

Berlin
Heidelberg
New York
Hong Kong
London
Milan
Paris
Tokyo

Physics and Astronomy

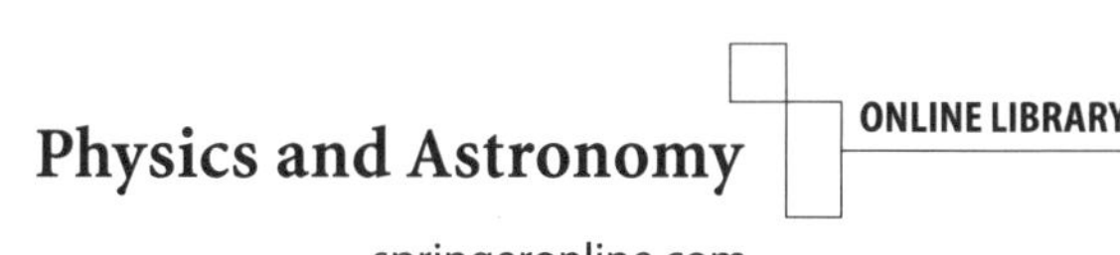

springeronline.com

Springer Tracts in Modern Physics

Springer Tracts in Modern Physics provides comprehensive and critical reviews of topics of current interest in physics. The following fields are emphasized: elementary particle physics, solid-state physics, complex systems, and fundamental astrophysics.

Suitable reviews of other fields can also be accepted. The editors encourage prospective authors to correspond with them in advance of submitting an article. For reviews of topics belonging to the above mentioned fields, they should address the responsible editor, otherwise the managing editor.
See also springeronline.com

Managing Editor

Gerhard Höhler

Institut für Theoretische Teilchenphysik
Universität Karlsruhe
Postfach 69 80
76128 Karlsruhe, Germany
Phone: +49 (7 21) 6 08 33 75
Fax: +49 (7 21) 37 07 26
Email: gerhard.hoehler@physik.uni-karlsruhe.de
www-ttp.physik.uni-karlsruhe.de/

Elementary Particle Physics, Editors

Johann H. Kühn

Institut für Theoretische Teilchenphysik
Universität Karlsruhe
Postfach 69 80
76128 Karlsruhe, Germany
Phone: +49 (7 21) 6 08 33 72
Fax: +49 (7 21) 37 07 26
Email: johann.kuehn@physik.uni-karlsruhe.de
www-ttp.physik.uni-karlsruhe.de/~jk

Thomas Müller

Institut für Experimentelle Kernphysik
Fakultät für Physik
Universität Karlsruhe
Postfach 69 80
76128 Karlsruhe, Germany
Phone: +49 (7 21) 6 08 35 24
Fax: +49 (7 21) 6 07 26 21
Email: thomas.muller@physik.uni-karlsruhe.de
www-ekp.physik.uni-karlsruhe.de

Fundamental Astrophysics, Editor

Joachim Trümper

Max-Planck-Institut für Extraterrestrische Physik
Postfach 16 03
85740 Garching, Germany
Phone: +49 (89) 32 99 35 59
Fax: +49 (89) 32 99 35 69
Email: jtrumper@mpe-garching.mpg.de
www.mpe-garching.mpg.de/index.html

Solid-State Physics, Editors

Andrei Ruckenstein
Editor for The Americas

Department of Physics and Astronomy
Rutgers, The State University of New Jersey
136 Frelinghuysen Road
Piscataway, NJ 08854-8019, USA
Phone: +1 (732) 445 43 29
Fax: +1 (732) 445-43 43
Email: andreir@physics.rutgers.edu
www.physics.rutgers.edu/people/pips/
Ruckenstein.html

Peter Wölfle

Institut für Theorie der Kondensierten Materie
Universität Karlsruhe
Postfach 69 80
76128 Karlsruhe, Germany
Phone: +49 (7 21) 6 08 35 90
Fax: +49 (7 21) 69 81 50
Email: woelfle@tkm.physik.uni-karlsruhe.de
www-tkm.physik.uni-karlsruhe.de

Complex Systems, Editor

Frank Steiner

Abteilung Theoretische Physik
Universität Ulm
Albert-Einstein-Allee 11
89069 Ulm, Germany
Phone: +49 (7 31) 5 02 29 10
Fax: +49 (7 31) 5 02 29 24
Email: frank.steiner@physik.uni-ulm.de
www.physik.uni-ulm.de/theo/qc/group.html

Frank Wissmann

Compton Scattering

Investigating the Structure of the Nucleon
with Real Photons

With 68 Figures and 13 Tables

Springer

Frank Wissmann
Physikalisch-Technische Bundesanstalt
Bundesallee 100
38116 Braunschweig, Germany
E-mail: Frank.Wissmann@ptb.de

Library of Congress Cataloging-in-Publication Data

Wissmann, Frank
Compton Scattering : investigating the structure of the nucleon with real photons / Frank Wissmann
p. cm. –(Springer tracts in modern physics, ISSN 0081-3869 ; v. 200)
Includes bibliographical references and index.
ISBN 3-540-40742-1 (acid-free paper)
1. Compton effect. 2. Scattering (Physics) I. Title. II. Series.

QC1.S797 vol.200
[QC794.6.S3]
539 s–dc22
[539.7′58]

20030503900

Physics and Astronomy Classification Scheme (PACS):
11.55.Fv, 11.55.Hx, 13.60.Fz, 14.20.Dh, 25.20.Dc

ISSN print edition: 0081-3869
ISSN electronic edition: 1615-0430
ISBN 3-540-40742-1 Springer-Verlag Berlin Heidelberg New York

Springer-Verlag is a part of Springer Science+Business Media

springeronline.com

Typesetting: printing from Author's data, using Springer LATEX macro package
Production: LE-TEXJelonek, Schmidt & Vöckler GbR, Leipzig
Cover concept: eStudio Calamar Steinen
Cover production: *design &production* GmbH, Heidelberg

Printed on acid-free paper   SPIN: 10946671   56/3141/YL   5 4 3 2 1 0

Preface

A very promising tool for studying the "low-energy structure" of the nucleon is elastic photon scattering (Compton scattering). It has the advantage of the initial and final states being identical. The weak coupling of the photon allows the use of perturbation theory for the description of the elementary interaction process. At low photon energies the scattering amplitude can be expanded in terms of the photon energy. The expansion parameters describe structure effects appearing at increasing photon energy. Such an expansion is valid up to the π-production threshold around 140 MeV. At higher energies, the technique of dispersion relations gives access to the reaction mechanisms which dominate Compton scattering. These are not limited to the excitation of the nucleon resonances. In addition, the excitation of non-resonant pion-nucleon states and the exchange of virtual mesons in the t-channel play a major role. All these features are related to the internal structure of the nucleon.

Compton scattering experiments are difficult because of the rather small cross sections ranging from about 10 nb/sr (low energies) to about 200 nb/sr (Δ resonance). In order to achieve reasonable statistical precision in the measured differential cross sections high intensity photon beams are required with sufficient energy resolution. The latter requirement is based on the strong energy dependence of the cross sections in certain energy regions. At the modern real-photon-beam laboratories photon intensities of the order $10^5 - 10^6\,\mathrm{s}^{-1}\mathrm{MeV}^{-1}$ are available. Therefore, a few hundred hours of beam time are necessary to ensure sufficient statistical precision for Compton scattering experiments. The data analysis following the experiments takes about 2 to 5 years depending on the complexity of the detection system.

This book emerges from my thesis [1] as a part of the "Habilitation" procedure at the Universität Göttingen (Germany). It summarizes the experiments[1] on Compton scattering from the proton and neutron performed at the real photon facility at the electron accelerator MAMI (Institut für Kernphysik, Universität Mainz, Germany). After a brief historical survey an introduction to the theoretical description of Compton scattering is given followed by the description of the experiments on the proton. The results

---

[1] Supported by Deutsche Forschungsgemeinschaft SFB 201, SFB 443 and by contracts Schu222, 436 RUS 113/510 and Wil198.

of these experiments are interpreted with the help of dispersion relations. The experiments investigating quasi-free Compton scattering from the nucleons bound in the deuteron are essential in the sense that any experiment on the neutron has to use nuclear targets, here the deuteron, since free neutron targets do not exist. The very first experiment on quasi-free scattering from the proton proved the correctness of the applied technique for extracting differential cross sections (for free proton scattering) from the quasi-free cross sections. Based on this success the experiment on the bound neutron was developed. The measured differential cross sections on Compton scattering from the neutron cover the entire Δ-resonance region and allowed a reliable determination of the electromagnetic polarizabilities of the neutron. After that series of Compton scattering experiments the electromagnetic as well as the backward spin polarizabilities have been determined experimentally. Therefore, a detailed investigation of the individual contributions to the polarizabilities of the proton and neutron and their prediction by chiral perturbation theory is possible.

Without the intense work of many people the experiments would not have been as successful as they are. Many people have contributed to the success of the experiments. The most important contribution came from the MAMI staff. They were responsible for the delivery of an always perfect electron beam. My gratitude also belongs to the group of Prof. Dr. M. Schumacher at the 2. Physikalisches Institut, Universität Göttingen and to the people of the A2 collaboration at the Institut für Kernphysik, Universität Mainz. Many thanks to Dr. A.I. L'vov (P. N. Lebedev Physical Institute, Moscow, Russia) for supplying me with the latest version of his computer code to calculate differential cross sections with his dispersion relation approach. The fruitful cooperation with Dr. M.I. Levchuk (B. I. Stepanov Institute of Physics, Minsk, Belorussia) made it possible to interpret the results on quasi-free Compton scattering from the proton and neutron bound in the deuteron.

I would like to express my gratitude to a very important person in my scientific life, Prof. Dr. M. Schumacher (Universität Göttingen). I appreciate his knowledge and understanding of physics as a teacher and scientist. I would like to thank him for all the freedom I was given during the experiments and for supporting all my activities at MAMI.

Braunschweig, September 2003   *Frank Wissmann*

References

1. F. Wissmann, *Compton Scattering and the Structure of the Nucleon*, Cuvillier, Göttingen, 2001

Contents

1 Introduction

The understanding of the basic constituents of nuclear matter, of how they interact and of what makes them build nuclei is still a challenging task for the scientific community. It covers a wide area of research, from nuclear spectroscopy at energies of a few MeV, to high–energy (particle) physics, utilizing particle beams with energies of a few hundreds of GeV. The link between these two extremes may be illustrated by means of the constituents of nuclei, i.e. protons and neutrons, which are called nucleons. For a detailed overview of the theoretical description of the structure of the nucleon and the quark models involved, the reader is referred to the book by Thomas and Weise [1].

Since a nucleon is built from point-like entities, namely the quarks [1], one may ask: "Does it have a spatial extension?" This is the topic of elastic electron scattering [2], where the nucleon structure is probed by the exchange of a virtual photon. Since photons couple only to electric charges and magnetic moments, this kind of reaction explores the charge and magnetic-moment distributions of an object, expressed in terms of form factors. In Fig. 1.1, the electric form factor G_E of the proton is plotted as a function of the four-momentum transfer squared [3]. The Q^2 dependence is well described by the dipole form factor (the dashed line in Fig. 1.1)

$$G_\mathrm{D} = \left(1 + a^2 Q^2\right)^{-2} , \tag{1.1}$$

where $a = 0.249$ fm [2]. At moderate Q^2, the electric form factor is the Fourier transform of the charge distribution. The rms charge radius can be derived from the slope of the electric form factor at very low Q^2. The results of experiments carried out in Mainz [4] in the late 1970s are still referred to when the rms radius of the proton is cited as [3, 5]

$$\sqrt{\langle r_\mathrm{e}^2\rangle_\mathrm{p}} = 0.862 \pm 0.012 \text{ fm} . \tag{1.2}$$

The inset in Fig. 1.1 shows the deduced charge distribution as a function of the radius r; it is an exponential function. The conclusion drawn from this is that the proton is neither a point-like particle nor a homogeneously charged sphere, it is a rather complex object.

From the above, one may deduce that *a particle which has a spatial extension and consists of constituents must have a resonant excitation spectrum.* This is indeed the case. In Fig. 1.2, a comparison between the total cross

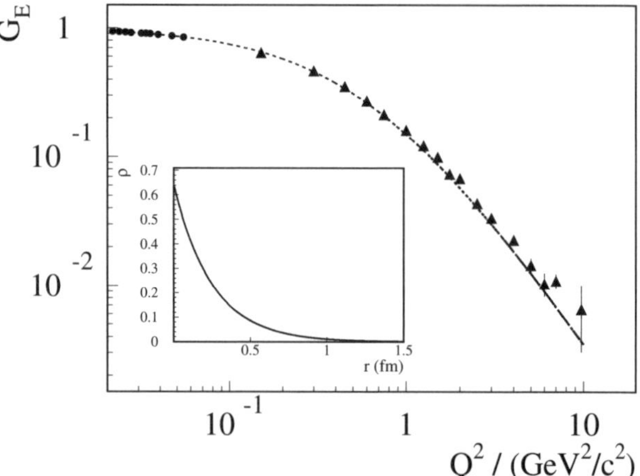

Fig. 1.1. The electric form factor of the proton as evaluated from the analysis of electron scattering experiments [3]. The data at very low Q^2 are from [4], from which the still valid rms radius of the proton, (1.2), has been determined. The dashed line is the dipole approximation, from which the charge distribution, shown in the inset, has been deduced

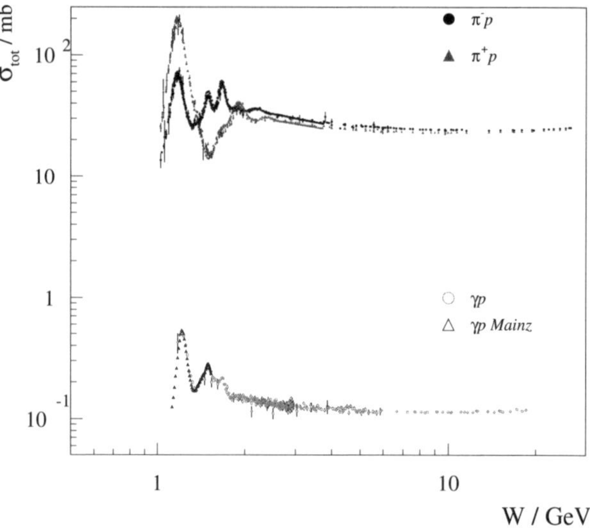

Fig. 1.2. The total cross section for pion-induced (π^+p, π^-p) and photon-induced (γp) reactions on the proton [6]. The first peak in the cross sections is due to the excitation of the Δ resonance. The difference between the π^+ and π^- cross sections exhibits the isospin dependence of the excitation mechanisms

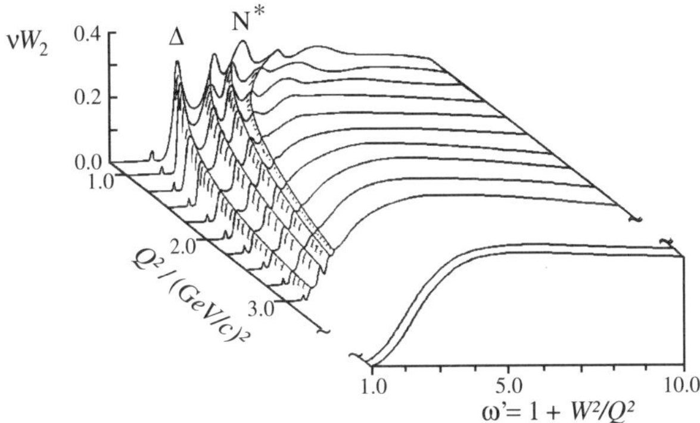

Fig. 1.3. The structure function W_2 as measured in electron scattering vs. the momentum transfer Q^2 and the scaling variable ω', where W is the total energy of the final system. From [7] with permission from Thomas Jefferson National Lab

sections obtained with hadronic and electromagnetic probes is shown. One common feature is the strong enhancement of the total cross section due to the excitation of the Δ resonance at 1232 MeV total energy. The isospin dependence of hadronic reactions is clearly visible. The huge difference in the absolute scale between the π- and γ-induced reactions displays the weak coupling of the photon. This is expressed by the coupling constant $e^2/4\pi = 1/137$ which is roughly 100 times smaller than the strong coupling of the π meson. At very high energies, both cross sections show a similar behavior. In this regime the photon reveals its "hadronic structure", which can be described by the *vector dominance model* [1]. According to this model, the photon is supposed to fluctuate into a ρ meson which has the same quantum numbers as the photon.

The transition from the constituent quarks to current quarks may be illustrated with the help of Fig. 1.3 [7]. The proton structure function νW_2 obtained from electron scattering is plotted versus the four-momentum transfer squared, $Q^2 = -q^2$, and the scaling variable ω'. The latter coincides, for large energy transfer, with $1/x$, where x is the Bjorken variable. For $Q^2 < 3$ $(GeV/c)^2$ and $\omega' < 5$, the excitation of resonant states is apparent. Outside this region, the structure function is very smooth and depends on the scaling variable ω' only. This region is known as the deep inelastic scattering region. There, the scaling behavior can be understood as incoherent scattering of the electrons from point-like quarks inside the proton. The link between the two regions is the domain of *medium-energy physics*, for which the energy range up to a few GeV covers the entire excitation spectrum of the proton. The reader should keep in mind that in electron scattering, the proton is probed by a virtual photon with $Q^2 > 0$. Thus, the energy and the momentum transfer can be selected independently.

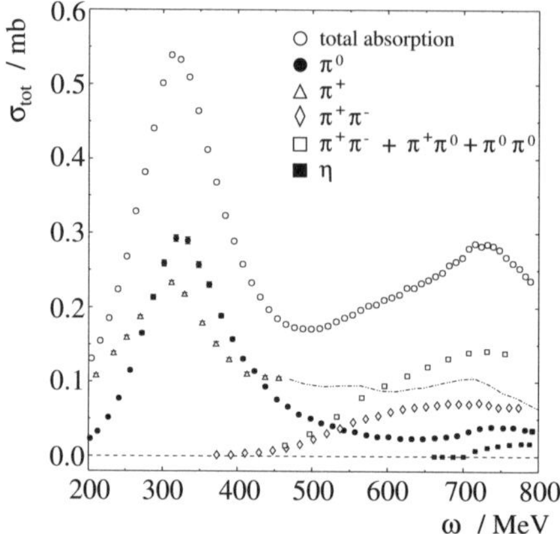

Fig. 1.4. The photoabsorption cross section of the proton from 200 MeV to 800 MeV as measured at the tagged photon beam at Mainz [8]

At the so-called photon point, where $Q^2 = 0$, the excitation mechanisms can be investigated with real photons. The absorption of a real photon is followed by the emission of secondary particles. These partial channels are shown in Fig. 1.4. The Δ resonance decays mainly via the emission of π mesons. The π^0 channel shows a strong resonant behavior, whereas the π^+ channel also has a large contribution from nonresonant π^+-photoproduction. The resonant and nonresonant contributions to the absorption cross section reveal the internal structure of the nucleon. The resonant excitation is related to the three-quark system, and the nonresonant part is due to the photon being coupled to virtual quark–antiquark systems, i.e. the pion cloud. Such a simple picture of the internal structure has to be investigated in more detail.

References

1. A. W. Thomas, W. Weise, *The Structure of the Nucleon*, Wiley-VCH, Berlin, 2001
2. D. Drechsel, M. M. Giannini, Rep. Prog. Phys. **52** (1989) 1083
3. R. C. Walker et al., Phys. Rev. D **49** (1994) 5671
4. G. G. Simon et al., Nucl. Phys. A **333** (1980) 381
5. P. Mergell et al., Nucl. Phys. A **596** (1996) 367
6. K. Hagiwara et al., Phys. Rev. D **66** (2002) 010001, *Review of Particle Physics*, Particle Data Group, `http://pdg.lbl.gov`
7. Thomas Jefferson National Laboratory, *Research Program at CEBAF: Report of the 1985 Summer Study Group*, Newport News, 1986; reprinted with permission from Thomas Jefferson National Laboratory
8. Institut für Kernphysik, Universität Mainz, *MAMI Jahresbericht 2000–2001*; reprinted with permission from Institut für Kernphysik, Universität Mainz

2 A Brief Historical Survey

Experiments on Compton scattering from the proton were first reported in the 1950s by Pugh et al. [1], Oxley et al. [2], Govorkov et al. [3] and Hyman et al. [4]. The photon energy range covered by these experiments extended from 30 MeV to 140 MeV. The first experimental results in the Δ resonance region were published in 1959 by Littauer et al. [5], followed by Gol'danski et al. [6], Bernardini et al. [7], DeWire et al. [8], Stiening et al. [9], Baranov et al. [10] and Gray et al. [11] in the 1960s. The photon beams were produced by electron beams from betatrons or synchrotrons hitting thin radiators and thus emitting bremsstrahlung. The principle used to detect the scattered photons was almost the same for all experiments. The scattered photons traversed a thick absorber, which reduced the electron background, followed by a telescope of plastic counters and converters. Pugh et al. used additionally a liquid scintillator to obtain information about the energy of the photon, which converted into an electron–positron pair.

At photon energies exceeding the π threshold, the recoiling proton is able to leave the target and can be detected in addition to the scattered photon. By making use of the kinematics, such a coincidence measurement not only helps to reduce electromagnetic background, but also helps to distinguish between photon–proton events caused by Compton scattering and events caused by π^0 photoproduction. As an example of how the data of these early experiments were recorded and analyzed, the article of Bernardini et al. [7] very accurately describes the electronic equipment. The following can be found:

> The three pulses were displayed on a Tektronix Model 517 A oscilloscope. A sweep with a speed of 50 ns/cm was used. It was triggered by the master coincidence output with the longer resolving time. (. . .)
>
> The oscilloscope traces were photographed with a DuMont Type 314 Oscillograph Record Camera. Eastman Kodak Linagraph Ortho Film ran at a constant speed in the camera, so that the film length indicated the time elapsed between the events.
>
> The processed films were projected on a screen from which the pulse heights and the timing of the three pulses were measured.

This example already demonstrates the very basic principle of modern data acquisition systems, except that the analog signals were transformed into digital information by the experimenters themselves. The result of this

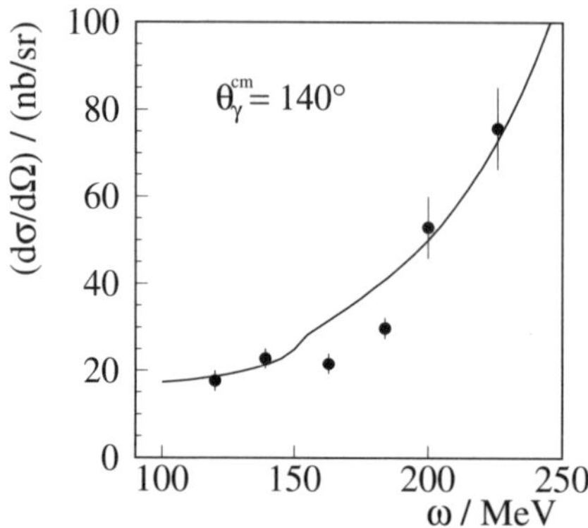

Fig. 2.1. The experimental results of Bernardini et al. [7] at a photon scattering angle of $\theta_\gamma^{cm} = 139.7°$. The *solid line* shows the most modern calculation in the framework of dispersion relations by L'vov et al. [12]

experiment (Fig. 2.1) shows remarkable agreement with the most modern theoretical calculation by L'vov et al. [12]. It should also be emphasized that the statistical uncertainties are rather small. It was concluded from these early experiments that the t-channel exchange of a neutral π meson, proposed by Low in 1954 [4, 7, 13], is of major importance [4, 7].

Compton scattering in the Δ-resonance region was continued in the 1970s by Genzel et al. [14] in Bonn. The outstanding feature of these experiments was that the energy dependence was measured over a wide angular range. For years these results were the basis of an intense discussion, because the data seemed to violate the unitarity bounds of Compton scattering. This puzzle has been solved by later experiments, discussed in this book. In 1980 Ishii et al. [15] performed some remarkable experiments which cover an energy range of 375 MeV up to 1150 MeV. Ishii et al. used a Pb-glass detector for the scattered photons and a high-resolution magnetic spectrometer to trace the recoiling protons.

The so-called "modern" experiments started in 1991 with those of Federspiel et al. [16], the first experiment which reliably determined the electromagnetic polarizabilities of the proton. All the subsequent experiments will be cited later in this book. In this brief overview, only the experiments covering an energy range up to the Δ resonance region have been considered. At higher energies, many more experiments have been performed. In Fig. 2.2, a kinematical overview of the experiments before 1990 is given. There are only few experiments in the Δ-resonance region from 200 MeV to 500 MeV. The challenge was to fill this gap, i.e. to perform experiments over a wide angular and energy range.

Today, two approaches to produce intense monoenergetic photon beams have emerged: (i) backscattering of laser light by high-energy electron beams,

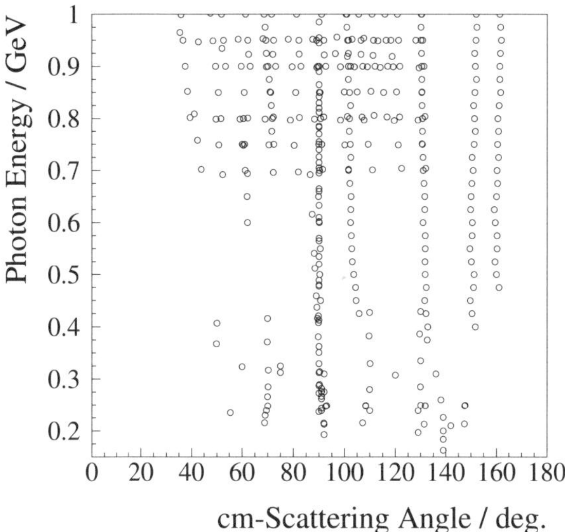

Fig. 2.2. This plot gives a kinematical overview of all experiments performed before 1990. Only experiments above the π threshold have been considered

and (ii) bremsstrahlung produced by electron beams hitting a thin radiator. Both techniques use an advantage of modern electron accelerators, compared with earlier machines, which is the high duty cycle of up to 100% and the low emittance of the electron beam. Monoenergetic photons are tagged by analyzing the electron which was hit by a laser photon or which emitted a bremsstrahlung photon, using a magnetic spectrometer. The difference between the electron energies before and after the reaction is attributed to the photon (photon tagging). The rate of tagged photons then is mainly limited by the counting rate of the electron detection system. With these techniques, a wide energy range may be covered with a reasonable photon energy resolution of 2 MeV to 5 MeV. This development has provided an important basis for the precise measurements of the energy dependence and angular dependence of the differential cross sections for Compton scattering described in the present book.

References

1. G. E. Pugh et al., Phys. Rev. **95** (1954) 590; G. E. Pugh et al., Phys. Rev. **105** (1957) 982
2. C. L. Oxley, V. L. Telegdi, Phys. Rev. **100** (1955) 435; C. L. Oxley, Phys. Rev. **110** (1958) 733
3. B. B. Govorkov et al., Sov. Phys. Doklady **1** (1956) 735
4. L. G. Hyman et al., Phys. Rev. Lett. **3** (1959) 93

5. R. M. Littauer et al., Bull. Am. Phys. Soc. **4** (1959) 253
6. V. I. Gol'danski et al., Sov. Phys. JETP **11** (1960) 1223
7. G. Bernardini et al., Il Nuovo Cimento **18** (1960) 1203
8. J. W. DeWire et al., Phys. Rev. **124** (1961) 909
9. R. F. Stiening et al., Phys. Rev. Lett. **10** (1963) 536
10. P. S. Baranov et al., Sov. J. Nucl. Phys. **3** (1966) 791
11. E. R. Gray, A. O. Hanson, Phy. Rev. **160** (1967) 1212
12. A. I. L'vov, V. A. Petrun'kin, M. Schumacher, Phys. Rev. C **55** (1997) 359
13. F. E. Low, Phys. Rev. **96** (1954) 1428
14. H. Genzel et al., Z. Phys. A **279** (1976) 399
15. T. Ishii et al., Nucl. Phys. B **165** (1980) 189
16. F. J. Federspiel et al., Phys. Rev. Lett. **67** (1991) 1511

3 Compton Scattering from the Proton

3.1 Low-Energy Expansion

3.1.1 Low-Energy Expansion to $\mathcal{O}(\omega\omega')$

At incident photon energies far below the π production threshold, i.e. $\omega \to 0$, photon scattering from the proton is described completely by the proton's static properties of mass and charge. In classical electrodynamics this corresponds to the scattering of an electromagnetic wave from a point particle with mass m and charge e (Thomson scattering). In quantum mechanics, this scattering process is described by the scattering amplitude, which in the case of Thomson scattering[1] is

$$f_{\mathrm{Th}} = -\frac{e^2}{m}\, \boldsymbol{\epsilon}' \cdot \boldsymbol{\epsilon} \,. \tag{3.1}$$

Here, $\boldsymbol{\epsilon}'$ and $\boldsymbol{\epsilon}$ are the polarization vectors of the incident and the scattered photon, and $e^2 = 1/137$.[2] With increasing photon energy, additional terms arise owing to the magnetic moment μ. These terms were calculated for the proton by Low [1], Gell-Mann and Goldberger [2, 3], and Klein [4] in fundamental articles. These articles give an expression for the expansion of the scattering amplitude up to terms linear in the photon energy ω. The existence of an anomalous magnetic moment suggests, however, that the proton might have an internal structure. By expanding the scattering amplitude up to terms of the order ω^2, Petrun'kin [5, 6] developed the complete scattering amplitude into the low-energy expansion

$$f_{\mathrm{Pet}} = -\frac{e^2}{m}\boldsymbol{\epsilon}'\boldsymbol{\epsilon} + \mathrm{i}\left(\omega' + \omega\right)\frac{e^2}{4m^2}\left(1 + 2\kappa\right)\boldsymbol{\sigma}\cdot\left(\boldsymbol{\epsilon}' \times \boldsymbol{\epsilon}\right)$$

$$-\mathrm{i}(\omega' + \omega)\frac{e^2}{4m^2}\left(1 + \kappa\right)^2 \boldsymbol{\sigma}\cdot\left[\left(\boldsymbol{n}' \times \boldsymbol{\epsilon}'\right) \times \left(\boldsymbol{n} \times \boldsymbol{\epsilon}\right)\right]$$

$$+\mathrm{i}\frac{e^2}{2m^2}\left(1 + \kappa\right)\left[\omega'\left(\boldsymbol{n}' \cdot \boldsymbol{\epsilon}\right)\boldsymbol{\sigma}\cdot\left(\boldsymbol{n}' \times \boldsymbol{\epsilon}'\right) - \omega\left(\boldsymbol{n} \cdot \boldsymbol{\epsilon}'\right)\boldsymbol{\sigma}\cdot\left(\boldsymbol{n} \times \boldsymbol{\epsilon}\right)\right]$$

---

[1] The reader should be aware of the fact that the Thomson amplitude of the neutron is equal to zero owing to the neutron's zero charge.

[2] Here and in the following Gaussian units, i.e. $e^2 = 1/137$, are used instead of Heaviside units, where $e^2/4\pi = 1/137$.

$$+\omega'\omega\frac{e^2}{4m^3}\left(2\kappa+\kappa^2\right)\boldsymbol{\epsilon}'\cdot\boldsymbol{\epsilon}$$

$$-\omega'\omega\frac{e^2}{4m^3}\left(1+\kappa\right)^2\left(\boldsymbol{n}'\times\boldsymbol{\epsilon}'\right)\cdot\left(\boldsymbol{n}\times\boldsymbol{\epsilon}\right)\left(\boldsymbol{n}'\cdot\boldsymbol{n}\right)$$

$$+\omega'\omega\frac{e^2}{4m^3}\left(\boldsymbol{n}'\times\boldsymbol{\epsilon}'\right)\cdot\left(\boldsymbol{n}\times\boldsymbol{\epsilon}\right)$$

$$+\omega'\omega[\alpha\boldsymbol{\epsilon}'\cdot\boldsymbol{\epsilon}+\beta\left(\boldsymbol{n}'\times\boldsymbol{\epsilon}'\right)\cdot\left(\boldsymbol{n}\times\boldsymbol{\epsilon}\right)]\,, \tag{3.2}$$

where κ is the anomalous magnetic moment of the proton,[3] α and β are the electric and magnetic polarizabilities,[4] respectively, and $\boldsymbol{n}$ and $\boldsymbol{n}'$ are the directions of the incident and the scattered photon; $\boldsymbol{n}'\cdot\boldsymbol{n}=\cos\theta_\gamma$, where θ_γ is the photon scattering angle in the laboratory system. The first term on the right-hand side of (3.2) is the Thomson amplitude f_{Th}. The next six terms describe the scattering due to the magnetic moment of the proton. Up to here the proton is assumed to be a point-like particle. The last term on the right-hand side of (3.2) is related to the internal structure of the proton and is expressed in terms of the electromagnetic polarizabilities α and β. The low-energy expansion given in (3.2) may be summarized as

$$f_{\mathrm{Pet}}=f_{\mathrm{Point}}+f_{\mathrm{Pol}}\,. \tag{3.3}$$

The necessity to expand the scattering amplitude up to the order ω^2 is now obvious. Calculating the differential cross section by squaring the amplitude, i.e.

$$\frac{\mathrm{d}\sigma}{\mathrm{d}\Omega}=|f|^2\,, \tag{3.4}$$

gives rise to ω^2 terms (i) due to the terms linear in ω or ω', and (ii) due to the interference of the Thomson amplitude f_{Th} and the terms proportional to $\omega\omega'$. Squaring the amplitude (3.2), averaging over the incident photon polarizations and proton spin states and summing over the final polarizations and spin states yields the differential cross section for the scattering of unpolarized photons from unpolarized protons [5, 6]:

$$\left(\frac{\mathrm{d}\sigma}{\mathrm{d}\Omega}\right)_{\mathrm{Pet}}=\left(\frac{\mathrm{d}\sigma}{\mathrm{d}\Omega}\right)_{\mathrm{Point}}$$

$$-\omega\omega'\left(\frac{\omega'}{\omega}\right)^2\frac{e^2}{m}\left[\frac{\alpha+\beta}{2}\left(1+z\right)^2+\frac{\alpha-\beta}{2}\left(1-z\right)^2\right]\,. \tag{3.5}$$

Here and in the following, $z=\cos\theta_\gamma$. The differential cross section of a point-like proton is given by

---

[3] The anomalous magnetic moment κ is defined as the difference between the magnetic moment μ of a particle with spin $1/2$ and charge $Q=e$ and the value for a Dirac particle: $\mu=(1+\kappa)\,e/2m$. For $Q=0$ (neutron), the magnetic moment is $\mu=\kappa\,e/2m$.

[4] The units used for the polarizabilities are given in Sect. A.1.

$$\left(\frac{d\sigma}{d\Omega}\right)_{\text{Point}} = \frac{1}{2}\left(\frac{e^2}{m}\right)^2\left(\frac{\omega'}{\omega}\right)^2 \times \{1 + z^2$$

$$+ \frac{\omega\omega'}{m^2}\left([1-z]^2 + a_0 + a_1 z + a_2 z^2\right)\} , \tag{3.6}$$

with the coefficients

$$a_0 = 2\kappa + \frac{9}{2}\kappa^2 + 3\kappa^3 + \frac{3}{4}\kappa^4 ,$$

$$a_1 = -4\kappa - 5\kappa^2 - 2\kappa^3 ,$$

$$a_2 = 2\kappa + \frac{1}{2}\kappa^2 - \kappa^3 - \frac{1}{4}\kappa^4 .$$

In some earlier articles, the low-energy expansion was developed in full detail. This means that the expression for the energy of the scattered photon,

$$\omega' = \frac{\omega}{1 + (\omega/m)\,(1-z)} , \tag{3.7}$$

was expanded in orders of (ω/m):

$$\left(\frac{\omega'}{\omega}\right) = 1 - \left(\frac{\omega}{m}\right)(1-z) + \left(\frac{\omega}{m}\right)^2(1-z)^2 - \left(\frac{\omega}{m}\right)^3(1-z)^3 + \cdots . \tag{3.8}$$

This led to quite lengthy formulas. To allow a better understanding, this expansion has been omitted in this book.

In the literature, one is often referred to the differential cross section evaluated by Powell, which is commonly known as the Powell cross section. In the appendix of his article [7], the differential cross section for scattering from a point-like particle with an anomalous magnetic moment κ has the form

$$\left(\frac{d\sigma}{d\Omega}\right)_{\text{Pow}} = \frac{1}{2}\left(\frac{e^2}{m}\right)^2\left(\frac{\omega'}{\omega}\right)^2\left\{\frac{\omega}{\omega'} + \frac{\omega'}{\omega} - (1-z^2)\right.$$

$$+\kappa\frac{2\omega\omega'}{m^2}(1-z)^2$$

$$+\kappa^2\frac{\omega\omega'}{m^2}[4(1-z) + \frac{1}{2}(1-z)^2]$$

$$+\kappa^3\frac{\omega\omega'}{m^2}[2(1-z) + (1-z^2)]$$

$$\left.+\kappa^4\frac{\omega\omega'}{2m^2}[1 + \frac{1}{2}(1-z^2)]\right\} . \tag{3.9}$$

The two expressions (3.6) and (3.9) are equivalent. For $\kappa = 0$, one obtains the Klein–Nishina cross section [8], the first line in (3.9), for scattering from a point-like charged particle with spin $1/2$. The relevance of κ and the structure of the proton, described by α and β as in (3.5), can be seen in Fig. 3.1 for

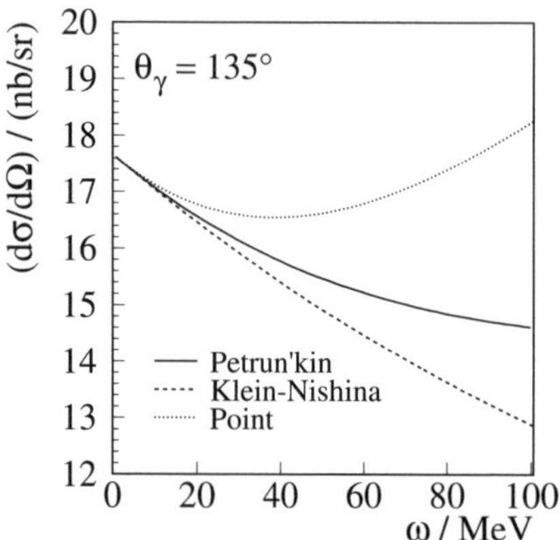

Fig. 3.1. The differential cross section for Compton scattering from a point-like proton in the low-energy expansion at $\theta_\gamma = 135°$ according to Petrun'kin [6] as given by (3.6) ("Point"). For $\kappa = 0$, the Klein–Nishina cross section for scattering from a point-like particle with spin $1/2$ is obtained ("Klein–Nishina"). The structure effects appear through the polarizabilities in (3.5). The calculation ("Petrun'kin") used the values $\alpha + \beta = 14.2$ and $\alpha - \beta = 10.0$

$\theta_\gamma = 135°$. The anomalous magnetic moment drastically increases the differential cross section compared with the Klein–Nishina cross section, because of the strong energy dependence of the κ terms. This is partially compensated by the interference term $f_{\mathrm{Th}} f_{\mathrm{Pol}}$ owing to the negative sign of the Thomson amplitude.

The low-energy expansion of the differential cross section (3.5) is valid at forward scattering angles [6]. At larger angles, one has to include the t-channel π^0 exchange [9, 10, 11], i.e. the π^0 pole contribution. According to Guiaşu et al. [10, 11], this leads to an additional term which changes (3.5) into

$$\left(\frac{\mathrm{d}\sigma}{\mathrm{d}\Omega}\right)_{\mathrm{Gui}} = \left(\frac{\mathrm{d}\sigma}{\mathrm{d}\Omega}\right)_{\mathrm{Pet}} + \frac{2}{m_\pi^2}\frac{\omega\omega'}{m^2}\left(\frac{\omega'}{\omega}\right)^2 (1-z)B_\pi(B_\pi + E)\,, \qquad (3.10)$$

where m_π is the π^0 mass and

$$B_\pi = \frac{m_\pi}{16\pi}g_{\pi\mathrm{NN}}F_{\pi^0\gamma\gamma}\frac{t}{m_\pi^2 - t}\,,$$

$$E = \frac{e^2}{m}\frac{m_\pi}{2}\left[1 - z + \kappa^2 + \kappa\,(3-z)\right]\,. \qquad (3.11)$$

Here, $g_{\pi\mathrm{NN}}$ and $F_{\pi^0\gamma\gamma}$ are the $\pi\mathrm{NN}$ and $\pi^0\gamma\gamma$ coupling constants, respectively, and t the four-momentum transfer squared: $t = (k - k')^2 = -2\omega\omega'(1 - z)$. Using the value $g_{\pi\mathrm{NN}}^2/4\pi = (13.75 \pm 0.15)$ [12] and the relation

$$F_{\pi^0\gamma\gamma}^2 = 64\pi\frac{\Gamma_{\pi^0\to\gamma\gamma}}{m_{\pi^0}^3}\,, \qquad (3.12)$$

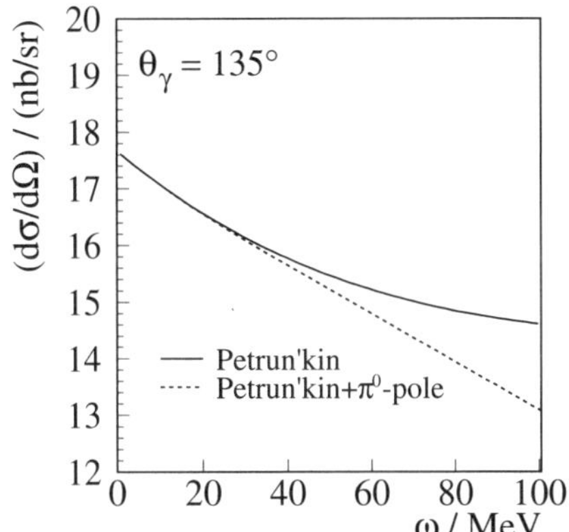

Fig. 3.2. The differential cross section for Compton scattering from the proton in the low-energy expansion at $\theta_\gamma = 135°$ according to [6] ("Petrun'kin") as given by (3.5). The polarizabilities were taken as $\alpha+\beta = 14.2$ and $\alpha-\beta = 10.0$. Including the contribution due to t-channel π^0 exchange ("π^0 pole") decreases the cross section at photon energies above 30 MeV

where $\Gamma_{\pi^0\to\gamma\gamma} = (7.83 \pm 0.56)$ eV [13] is the $\pi^0 \to \gamma\gamma$ decay width, the product of the couplings is

$$g_{\pi\mathrm{NN}}F_{\pi^0\gamma\gamma} = (-0.333 \pm 0.012)\,\mathrm{GeV}^{-1}\,. \tag{3.13}$$

The sign of the product $g_{\pi\mathrm{NN}}F_{\pi^0\gamma\gamma}$ is ambiguous when (3.12) is used, and in the following we assume $g_{\pi\mathrm{NN}}F_{\pi^0\gamma\gamma} < 0$, in accordance with arguments based on the axial anomaly [14]. As will be shown later, the sign may also be determined from the experimental scattering cross sections. From Fig. 3.2, it becomes obvious that the t-channel contribution cannot be neglected at far backward angles, not even at rather moderate incident photon energies. It strongly decreases the cross section. The importance of such processes, i.e. t-channel exchange of mesons coupling to two photons, is a special feature of a 2γ process such as Compton scattering.

3.1.2 Energy Expansion to $\mathcal{O}((\omega\omega')^2)$

In the previous subsection the expression for low-energy Compton scattering has been given to the order $\mathcal{O}(\omega\omega')$, including the contribution from the t-channel π^0 exchange. Higher-order terms were first introduced by Guiaşu and Radescu [10, 11]. One has to take into account all possible interference terms when deriving the cross section up the order $\mathcal{O}((\omega\omega')^2)$ from the scattering amplitudes. The expression obtained [10, 11] extends the cross section of (3.10):

$$\left(\frac{\mathrm{d}\sigma}{\mathrm{d}\Omega}\right)_{\omega^2\omega'^2} = \left(\frac{\mathrm{d}\sigma}{\mathrm{d}\Omega}\right)_{\mathrm{Gui}} - \frac{1}{4m^2}\frac{e^2}{m}\left(\frac{\omega'}{\omega}\right)^2 \omega^2\omega'^2$$

$$\times\left\{\frac{(1-z)^2}{2}[2 - 2\kappa - \kappa^2 - z\,(2 + 2\kappa + \kappa^2)](\alpha - \beta)\right.$$

$$-\frac{(1+z)^2}{2}(1-z)(2\kappa + \kappa^2)(\alpha + \beta)$$

$$-m^2\frac{m}{e^2}(1-z)^2(\alpha - \beta)^2$$

$$\left.-m^2\frac{m}{e^2}(1+z)^2(\alpha + \beta)^2\right\}$$

$$-\frac{e^2}{m}\left(\frac{\omega'}{\omega}\right)^2 \omega^2\omega'^2\frac{1}{64m^5}\left(\frac{m}{m_{\pi^0}}\right)^4$$

$$\times\left(\bar{A} + \bar{B}z + \bar{C}z^2 + \bar{D}z^3\right)\,. \tag{3.14}$$

Here, the next-order constants related to the internal structure of the proton, i.e. $\bar{A}$, $\bar{B}$, $\bar{C}$ and $\bar{D}$, are rather difficult to interpret in terms of physical quantities. The authors of [10] speculate about the nature of the new coefficients. In deriving an expression for the scattering of electromagnetic waves from a proton based on classical electrodynamics, these authors found a connection between the coefficient $\bar{D}$ and the quadrupole polarizability of the proton.

The most recent expansion of the unpolarized Compton scattering cross section was evaluated by Babusci et al. [15]. In terms of the invariant functions W_{ij}, this expansion has the following form:

$$\left(\frac{\mathrm{d}\sigma}{\mathrm{d}\Omega}\right)_{\omega^2\omega'^2} = \frac{1}{(8\pi m)^2}\left(\frac{\omega'}{\omega}\right)^2 (W^{\mathrm{B}}_{00} + W^{\mathrm{NB}}_{00}) \tag{3.15}$$

$$= \frac{1}{(8\pi m)^2}\left(\frac{\omega'}{\omega}\right)^2 \left(W^{\mathrm{B}}_{00} + \omega\omega'U^{(2)}_{00} + (\omega\omega')^2 U^{(4)}_{00}\right)\,. \tag{3.16}$$

The function W_{00} is split into the Born contribution (B) and the non-Born contribution (NB). For the latter, the energy expansion is given by the two functions $U^{(2)}_{00}$ and $U^{(4)}_{00}$:

$$\frac{1}{(8\pi m)^2}W^{\mathrm{B}}_{00} = \frac{1}{2}\left(\frac{e^2}{m}\right)^2\left\{1 + z^2 + \frac{\omega\omega'}{4m^2}\left[4(1 + 2\kappa)(1-z)^2\right.\right.$$

$$+2(9 - 10z + z^2)\kappa^2 + 4(3 - 2z - z^2)\kappa^3$$

$$\left.\left.+(3 - z^2)\kappa^4\right]\right\}\,, \tag{3.17}$$

$$\frac{1}{(8\pi m)^2}U^{(2)}_{00} = -\frac{e^2}{m}\left\{(1 + z^2)\alpha + 2z\beta\right\}\,, \tag{3.18}$$

$$\frac{1}{(8\pi m)^2}U^{(4)}_{00} = \frac{1 + z^2}{2}(\alpha^2 + \beta^2) + 2z\alpha\beta + \frac{e^2}{m}\frac{1 - z}{4m^2}$$

$$\times\left\{(1 + z)(\kappa^2 + 2\kappa)(\alpha + \beta z) + 4\alpha z + 2\beta(1 + z^2)\right\}$$

$$-\frac{e^2}{m}\left\{(1+z^2)\alpha_{\mathrm{E}\nu} + 2\beta_{\mathrm{M}\nu}z + \frac{z^3}{6}\alpha_{\mathrm{E}2} + \frac{3z^2-1}{12}\beta_{\mathrm{M}2}\right\}$$

$$+\frac{1}{2m}\frac{e^2}{m}\,P(z)\,. \tag{3.19}$$

The terms of (3.17) and (3.18) may be rewritten, and the equality with (3.5) becomes obvious.

The functions $U_{00}^{(2)}$ and $U_{00}^{(4)}$ display (i) the interference between the amplitudes f_{Point} and f_{Pol} of the low-energy expansion in (3.3), (ii) the square of the amplitude f_{Pol}, and (iii) additional terms arising from the expansion indicated by the new structure constants $\alpha_{\mathrm{E}\nu}$, $\beta_{\mathrm{M}\nu}$, which are corrections to the dipole polarizabilities α and β, and $\alpha_{\mathrm{E}2}$ and $\beta_{\mathrm{M}2}$, called the electric and magnetic quadrupole polarizabilities, respectively. Rearranging the third line in (3.19) into powers of z confirms the nature of the constant $\bar{D}$ in the formalism of Guiaşu [10] (see (3.14)), as the quadrupole polarizability $\alpha_{\mathrm{E}2}$. All coefficients labeled as α_x and β_x are related to the electromagnetic structure of the proton. The function $P(z)$, in contrast, describes the proton spin structure expressed by the spin polarizabilities $\gamma_{\mathrm{E}1}$, $\gamma_{\mathrm{M}1}$, $\gamma_{\mathrm{E}2}$ and $\gamma_{\mathrm{M}2}$. These coefficients appear only in the spin-dependent part of the scattering amplitude [15]. The function $P(z)$ reads as

$$\begin{aligned}
P(z) = \;& \gamma_{\mathrm{E}1}\left\{1 + 2z - 3z^2 - 2\kappa(1-z)^2 + 2\kappa^2 z\right\} \\
&+\gamma_{\mathrm{M}1}\left\{(1+2\kappa)(3 - 2z - z^2) + \kappa^2(3 - z^3)\right\} \\
&+\gamma_{\mathrm{E}2}\left\{-1 + 3z^2 - 2z^3 - 2\kappa(1 + z - 3z^2 + z^3)\right. \\
&\qquad\qquad \left.+\kappa^2(3z^2 - 1)\right\} \\
&+\gamma_{\mathrm{M}2}\left\{-(1-z)^2 + 4\kappa z(1-z) + 2\kappa^2 z\right\}\,.
\end{aligned} \tag{3.20}$$

The spin polarizabilities in (3.20) are linear combinations of the γ_i introduced by Ragusa [16]:

$$\begin{aligned}
\gamma_{\mathrm{E}1} &= -\gamma_1 - \gamma_3\,, \\
\gamma_{\mathrm{M}1} &= \gamma_4\,, \\
\gamma_{\mathrm{E}2} &= \gamma_2 + \gamma_4\,, \\
\gamma_{\mathrm{M}2} &= \gamma_3\,.
\end{aligned} \tag{3.21}$$

The physical meaning of the $\gamma_{\mathcal{M}L}$, where $\mathcal{M} = \mathrm{E}$, M and $L = 1, 2$, is the following. The quantities $\gamma_{\mathrm{E}1}$ and $\gamma_{\mathrm{M}1}$ are related to the spin dependence of the dipole transitions of the scattering process, i.e. E1 $\rightarrow$ E1 and M1 $\rightarrow$ M1, respectively. The dipole–quadrupole transition, i.e. M1 $\rightarrow$ E2 and E1 $\rightarrow$ M2, are related to $\gamma_{\mathrm{E}2}$ and $\gamma_{\mathrm{M}2}$.

The importance of the t-channel π^0 exchange contribution to the scattering process was outlined in [9, 10, 11] and this contribution was introduced in (3.10). To consider this amplitude in the $\mathcal{O}((\omega\omega')^2)$ expansion, we make the following replacements in (3.20) in accordance with [15]:

Table 3.1. The higher-order parameters used for the calculation of the cross section shown in Fig. 3.3, performed according to (3.15). All values were determined with the help of dispersion relations (see Sect. 3.2 and Chap. 6). The units are given in Sect A.1

$\alpha_{E\nu}$	$\beta_{M\nu}$	α_{E2}	β_{M2}	γ_{E1}	γ_{M1}	γ_{E2}	γ_{M2}
-3.7	9.1	28.7	-22.6	7.1	-8.1	-8.8	11.1

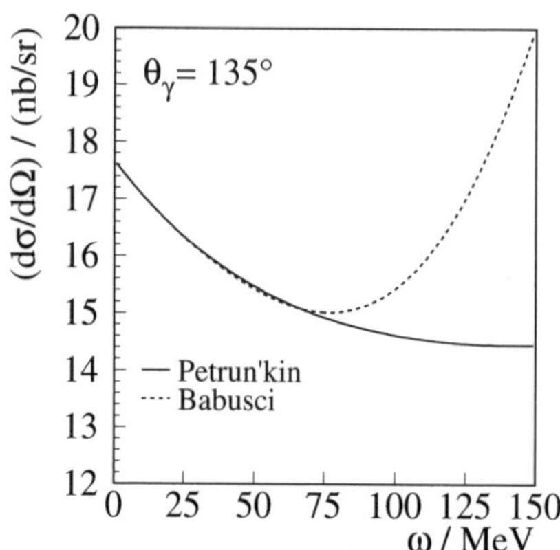

Fig. 3.3. The differential cross section for Compton scattering from the proton in the low-energy expansion up to the order $\mathcal{O}((\omega\omega')^2)$ at $\theta_\gamma = 135°$ according to [15] ("Babusci") as given in (3.15). The parameters used are tabulated in Table 3.1. For comparison, the low-energy expansion up to the order $\mathcal{O}(\omega\omega')$ according to [6] ("Petrun'kin") (see (3.5)), is also shown. The dipole polarizabilities were taken as $\alpha + \beta = 14.2$ and $\alpha - \beta = 10.0$

$$\gamma_{E1} \rightarrow \gamma_{E1} + \frac{g_{\pi NN} F_{\pi^0 \gamma\gamma}}{m_{\pi^0}^2} \frac{t}{t - m_{\pi^0}^2} \,,$$

$$\gamma_{M1} \rightarrow \gamma_{M1} - \frac{g_{\pi NN} F_{\pi^0 \gamma\gamma}}{m_{\pi^0}^2} \frac{t}{t - m_{\pi^0}^2} \,,$$

$$\gamma_{E2} \rightarrow \gamma_{E2} - \frac{g_{\pi NN} F_{\pi^0 \gamma\gamma}}{m_{\pi^0}^2} \frac{t}{t - m_{\pi^0}^2} \,,$$

$$\gamma_{M2} \rightarrow \gamma_{M2} + \frac{g_{\pi NN} F_{\pi^0 \gamma\gamma}}{m_{\pi^0}^2} \frac{t}{t - m_{\pi^0}^2} \,. \tag{3.22}$$

Now it becomes obvious that the t-channel exchange of a pseudoscalar particle, i.e. the π^0 meson, strongly dominaties the spin dependence of the scattering process. This will be described in more detail in Sect. 3.2. Here again, the relative sign of the coupling constants $g_{\pi NN} F_{\pi^0 \gamma\gamma}$ may be determined from the experimental scattering cross sections. Altogether, there are now 10 parameters to describe Compton scattering from the proton at low photon energies which are very difficult to disentangle. But, compared with

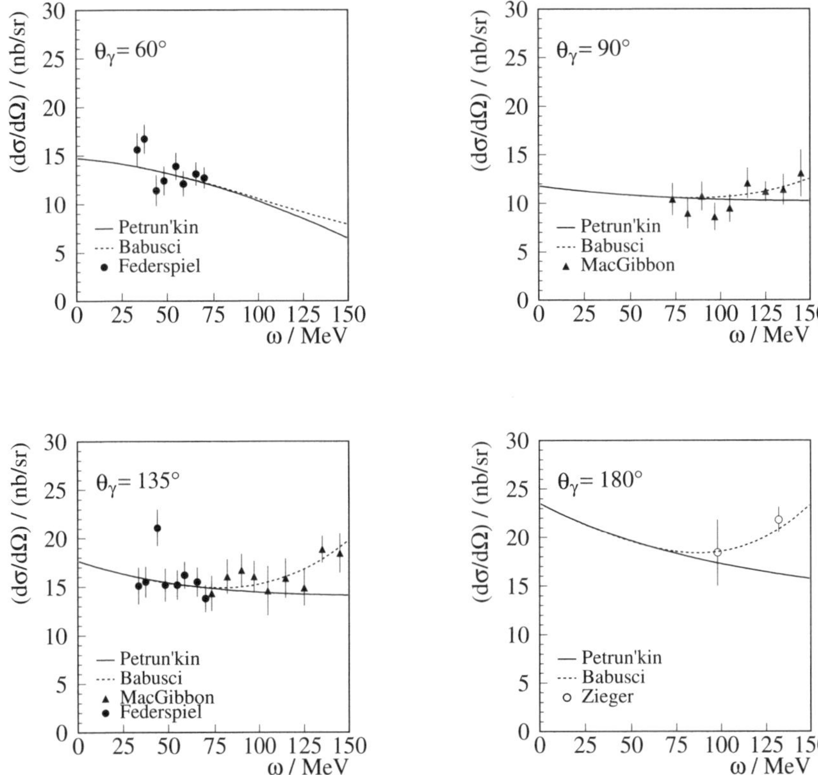

Fig. 3.4. Comparison of the low-energy expansions according to Petrun'kin [6] and Babusci et al. [15] and the differential cross sections measured by Federspiel et al. [17], MacGibbon et al. [18] and Zieger et al. [19]. The calculations used $\alpha + \beta = 14.2$ and $\alpha - \beta = 10.5$, and additional values as given in Table 3.1 for the Babusci expansion

the expansion given in [10, 11], we are now in a situation where some of the parameters may be fixed with the help of dispersion relations as it will be described in Sect. 3.2.

Using this theoretical description, the differential cross section of (3.15) has been calculated with the parameters given in Table 3.1 and is plotted in Fig. 3.3. The influence of the higher-order terms in the low-energy expansion up to the order $\mathcal{O}((\omega\omega')^2)$ becomes significant at photon energies above 80 MeV. This fact is, of course, not new it just displays the onset of the excitation of the Δ resonance. A comparison with experimental results (Fig. 3.4) shows the importance of using the $\mathcal{O}((\omega\omega')^2)$ expansion for incident photon energies up to the π production threshold.

The electromagnetic polarizabilities to be used in the low-energy expansions to the orders $\mathcal{O}(\omega\omega')$ and $\mathcal{O}((\omega\omega')^2)$ can be obtained by fitting the $\mathcal{O}((\omega\omega')^2)$ expansion to the experimental data of Federspiel et al. [17], MacGibbon et al. [18] and Zieger et al. [19], using the higher order polarizabilities as tabulated in Table 3.1. In addition, α and β are constrained by the Baldin sum rule $\alpha + \beta = 14.2 \pm 0.3$ ((3.41) and [20, 21]). Leaving $\alpha - \beta$ as a parameter results in

$$\alpha - \beta = 10.5 \pm 1.1 \tag{3.23}$$

in the units given in Sect. A.1 The uncertainty in (3.23) is the combined statistical and systematic uncertainty. Using $\alpha+\beta$ as an additional parameter, the values

$$\begin{aligned} \alpha + \beta &= 13.9 \pm 2.7 \, , \\ \alpha - \beta &= 10.6 \pm 1.2 \, , \end{aligned} \tag{3.24}$$

have been obtained. One remark has to be added concerning the fitting procedure. As explained in [18], the experiments by MacGibbon et al. were performed with tagged ($\omega < 100$ MeV) and untagged ($\omega > 100$ MeV) photons. Therefore, the differential cross sections for photon energies greater than 100 MeV (untagged) are strongly correlated, and the fitting of these data points was done in the same way as that described by MacGibbon et al. [18].

3.1.3 Forward Scattering and Sum Rules

The fundamental principles governing Compton scattering are unitarity and causality. The causality requirements, formulated in quantum field theory and applied to Compton scattering, are discussed in the fundamental article by Gell-Mann et al. [2]. Causality means that waves do not propagate faster than the speed of light. Applied to the scattering process, this implies that scattering can occur only after the incident wave has reached the target. The scattered wave then always propagates behind the original wave and not in front of it. Unitarity is related to probability conservation, which, when applied to quantum mechanical operators, requires the unitarity of the scattering matrix. Two fundamental relations follow from these two principles. The first is the *optical theorem* [22, 23],

$$\mathrm{Im}\, f(\omega) = \frac{\omega}{4\pi}\sigma_{\mathrm{tot}} \, , \tag{3.25}$$

which relates the imaginary part of the forward scattering amplitude to the total photoabsorption cross section. The second is the Kramers–Kronig dispersion relation[5] [24, 25],

$$\mathrm{Re}\, f(\omega) = \mathrm{Re}\, f(0) + \frac{\omega^2}{2\pi^2}\mathcal{P} \int_{\omega_{\mathrm{thr}}}^{\infty} \frac{\sigma_{\mathrm{tot}}(\omega')}{\omega'^2 - \omega^2}\,\mathrm{d}\omega' \, , \tag{3.26}$$

---

[5] The causality requirements used to deduce the Kramers–Kronig dispersion relation are quite different from those used to describe the scattering of elementary particles [2].

which allows one to calculate the real part of the forward scattering amplitude.[6] In (3.26), the physical threshold ω_{thr} has been introduced as the lower integration limit, since in the case of the nucleon the absorption cross section is zero below the π production threshold. This formalism allows one to calculate the scattering cross section at $\theta_\gamma = 0°$ at any energy on the basis of the measured total absorption cross section if the latter is known over a sufficient wide energy range. Even at photon energies below the π production threshold, the scattering process depends on the internal structure of the particle that is evident at higher energies. This is expressed in (3.26) by the absorption cross section in the integrand. In order to investigate this in more detail, the low-energy expansion of the scattering amplitude and its connection to the dispersion relations have to be evaluated.

Considering the extreme case of forward scattering, the Compton scattering amplitude valid at all energies has a simple form [2, 3]:

$$f(\omega, \theta = 0) = f_1(\omega)\boldsymbol{\epsilon}' \cdot \boldsymbol{\epsilon} + \mathrm{i}f_2(\omega)\boldsymbol{\sigma} \cdot (\boldsymbol{\epsilon}' \times \boldsymbol{\epsilon}) \ . \tag{3.27}$$

Here, f_1 describes the spin-independent and f_2 the spin-dependent part of the scattering amplitude. If we use the expression given by Petrun'kin [6], (3.2), and the expression given by Babusci et al. [15] for higher orders in ω, the low-energy expansions of f_1 and f_2 are

$$f_1(\omega) = -\frac{e^2}{m} + (\alpha + \beta)\,\omega^2 + (\alpha_2 + \beta_2)\,\omega^4 + \mathcal{O}(\omega^6) \ , \tag{3.28}$$

$$f_2(\omega) = -\frac{e^2\kappa^2}{2m^2}\omega + \gamma\omega^3 + \mathcal{O}(\omega^5) \ , \tag{3.29}$$

where $\alpha_2 + \beta_2 = \alpha_{\mathrm{E}\nu} + \beta_{\mathrm{M}\nu} + (1/12)\alpha_{\mathrm{E}2} + (1/12)\beta_{\mathrm{M}2}$, and the forward spin polarizability γ is defined as $\gamma = -\gamma_{\mathrm{E}1} - \gamma_{\mathrm{M}1} - \gamma_{\mathrm{E}2} - \gamma_{\mathrm{M}2}$ by the spin polarizabilities introduced in (3.20). The meaning of the quantities α_2 and β_2 will be explained in more detail in Sect. 3.2.

To show how these two amplitudes are related to each other, the scattering process has to be described in terms of helicity amplitudes. In the center-of-mass (cm) system, the helicity of a particle is the projection of its spin onto the momentum direction. Therefore, the helicity of the photon can have the values $\lambda_\gamma = \pm 1$, whereas the values for the proton are $\lambda_\mathrm{p} = \pm 1/2$. The total helicity of the γ–p system, $\lambda = |\lambda_\gamma - \lambda_\mathrm{p}|$, can thus take the values $1/2$ and $3/2$. In other words, the incident photon is circularly and the proton longitudinally polarized, and the polarization vectors are oriented parallel (helicity state $3/2$) or antiparallel (helicity state $1/2$) to each other. The polarization vectors of the photon are given by the right-handed ($\lambda_\gamma = +1$) and left-handed ($\lambda_\gamma = -1$) polarization vectors $\boldsymbol{\epsilon}_{\lambda_\gamma}$ [27],

---

[6] Concerning the formalism of dispersion relations the reader is referred to the recently published article by Drechsel et al. [26].

$$\epsilon_{+1} = -\frac{1}{\sqrt{2}} \left(\epsilon_x + i\epsilon_y \right) , \tag{3.30}$$

$$\epsilon_{-1} = \frac{1}{\sqrt{2}} \left(\epsilon_x - i\epsilon_y \right) , \tag{3.31}$$

which now become complex. If we look only at the case of $\lambda_\gamma = +1$ we can substitute

$$\epsilon' \cdot \epsilon \;\rightarrow\; \epsilon_{\lambda'}'^* \cdot \epsilon_\lambda \;=\; \begin{cases} 1 & \text{for } \lambda' = +1 \\ 0 & \text{for } \lambda' = -1 \end{cases} ,$$

$$\epsilon' \times \epsilon \rightarrow \epsilon_{\lambda'}'^* \times \epsilon_\lambda \;=\; \begin{cases} i e_z & \text{for } \lambda' = +1 \\ 0 & \text{for } \lambda' = -1 \end{cases} . \tag{3.32}$$

Hence, (3.27) may be rewritten by setting $\boldsymbol{\sigma} e_z = \sigma_z = \pm 1$, as

$$f_{1/2}(\omega) = f_1(\omega) + f_2(\omega) , \tag{3.33}$$
$$f_{3/2}(\omega) = f_1(\omega) - f_2(\omega) , \tag{3.34}$$

from which it follows that

$$f_1(\omega) = \frac{1}{2} \left(f_{1/2}(\omega) + f_{3/2}(\omega) \right) , \tag{3.35}$$

$$f_2(\omega) = \frac{1}{2} \left(f_{1/2}(\omega) - f_{3/2}(\omega) \right) . \tag{3.36}$$

With the help of the *optical theorem*, the imaginary parts of the amplitudes f_1 and f_2 are given by:

$$\mathrm{Im} f_1 (\omega) = \frac{\omega}{8\pi} \left(\sigma_{1/2}(\omega) + \sigma_{3/2}(\omega) \right) = \frac{\omega}{4\pi} \sigma_{\mathrm{tot}}(\omega) , \tag{3.37}$$

$$\mathrm{Im} f_2 (\omega) = \frac{\omega}{8\pi} \left(\sigma_{1/2}(\omega) - \sigma_{3/2}(\omega) \right) . \tag{3.38}$$

Applying the formalism of dispersion relations [26, 28, 29] to the amplitudes f_1 and f_2 with the restriction $\omega \ll \omega_{\mathrm{thr}}$, we obtain the following low-energy expansions of the dispersion relations [28]:

$$\mathrm{Re} \, f_1(\omega) = \mathrm{Re} \, f_1(0) + \frac{\omega^2}{2\pi^2} \int_{\omega_{\mathrm{thr}}}^{\infty} \frac{\sigma_{\mathrm{tot}}(\omega')}{\omega'^2} d\omega' , \tag{3.39}$$

$$\mathrm{Re} \, f_2(\omega) = \mathrm{Re} \, f_2(0) + \frac{\omega}{4\pi^2} \int_{\omega_{\mathrm{thr}}}^{\infty} \frac{\sigma_{1/2}(\omega') - \sigma_{3/2}(\omega')}{\omega'} d\omega' . \tag{3.40}$$

To remind the reader, at these photon energies the absorption cross section vanishes and the scattering is described by the real part of the amplitude

only. A comparison between (3.28) and (3.39) leads to the Baldin sum rule
[30],

$$\alpha + \beta = \frac{1}{2\pi^2} \int_{\omega_{\mathrm{thr}}}^{\infty} \frac{\sigma_{\mathrm{tot}}(\omega')}{\omega'^2} \mathrm{d}\omega' \, , \tag{3.41}$$

if $\mathrm{Re}\, f_1(0) = -e^2/m$. The conclusion drawn is that for the spin-independent
part of the scattering amplitude, a once-subtracted dispersion relation ap-
plies. From this relation, it becomes clear that the electromagnetic polarizabilities α and β are not just expansion parameters of the scattering amplitude. This expression reveals that the internal structure, i.e. the excitation
spectrum, plays a major role even at low photon energies.

One more fundamental sum rule arises from comparing (3.29) and (3.40),

$$-\frac{e^2\kappa^2}{2m^2} = \frac{1}{4\pi^2} \int_{\omega_{\mathrm{thr}}}^{\infty} \frac{\sigma_{1/2}(\omega') - \sigma_{3/2}(\omega')}{\omega'} \mathrm{d}\omega' \, , \tag{3.42}$$

if $\mathrm{Re}\, f_2(0) = 0$ [28], from which it follows that the spin-dependent part may
be described by a nonsubtracted dispersion relation. Equation (3.42) is called
the Gerasimov–Drell–Hearn (GDH) sum rule [31]. It gives a direct relation
between the anomalous magnetic moment κ and the helicity dependence of
the photoabsorption cross section.

From the higher-order terms, additional important sum rules can be deduced:

$$\alpha_2 + \beta_2 = \frac{1}{2\pi^2} \int_{\omega_{\mathrm{thr}}}^{\infty} \frac{\sigma_{\mathrm{tot}}(\omega')}{\omega'^4} \mathrm{d}\omega' \, , \tag{3.43}$$

$$\gamma = \frac{1}{4\pi^2} \int_{\omega_{\mathrm{thr}}}^{\infty} \frac{\sigma_{1/2}(\omega') - \sigma_{3/2}(\omega')}{\omega'^3} \mathrm{d}\omega' \, . \tag{3.44}$$

Equation (3.43) is referred to as the quadrupole sum rule and (3.44) as the
forward spin polarizability sum rule. Owing to the ω^{-3} and ω^{-4} weighting of
the absorption cross sections in the integrands, the influence of the threshold region is enhanced as compared with the Baldin and GDH sum rules.
Experimentally, this energy region is rather difficult to access. The reaction
$\mathrm{p}(\gamma, \pi^0)\mathrm{p}$ can be measured down to the threshold by detecting the produced
π^0 via its 2γ decay. In the absorption channel $\mathrm{p}(\gamma, \pi^+)\mathrm{n}$, the difficulty is that
most of the emitted π^+ mesons will be absorbed within the target. Thus,
the recoiling neutron has to be detected. The disadvantage of this technique
is the low detection efficiency of neutron detectors for low-energy neutrons.
Therefore, one may use a different method to extrapolate the cross section
down to the threshold, as will be explained in the next section.

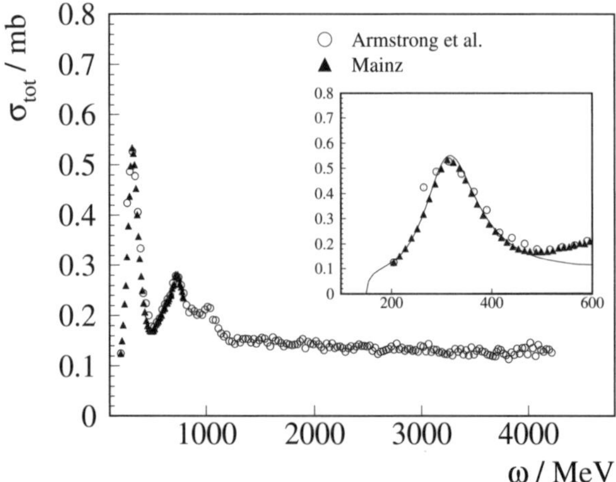

Fig. 3.5. The total photoabsorption cross section of the proton as measured by Armstrong et al. [32] and at Mainz [33]. The inset enhances the energy region up to 600 MeV. The prediction for single-π photoproduction, as calculated from the SAID partial-wave analysis of Arndt et al. [34], solution SM99K (*solid line* in the inset), shows that up to about 450 MeV, the absorption cross section is predominantly determined by this process

3.1.4 Reevaluation of the Sum Rules

The photoabsorption cross section of the proton (Fig. 3.5) was measured by Armstrong et al. [32] with tagged photons in the energy range 265 MeV to 4215 MeV in the early 1970s. In Fig. 3.5, the results are compared with the very precise data obtained recently in the energy range 200 MeV to 800 MeV [33] with the tagged photon beam in Mainz.

In the Δ-resonance region, there is a slight difference between the two measurements. The prediction of the SAID partial-wave analysis (Sect. A.2) by Arndt et al. [34], using the solution SM99K (solid line in the inset of Fig. 3.5), exhibits a rather good agreement with the Mainz data within the Δ-resonance region even though it is not perfect.

To evaluate the Baldin sum rule (3.41), the absorption cross section was divided into the following energy intervals:

$$\mathcal{I}_1 : \omega \in [145, 200) \ ,$$
$$\mathcal{I}_2 : \omega \in [200, 2000) \ ,$$
$$\mathcal{I}_3 : \omega \in [2000, \infty) \ . \tag{3.45}$$

In the interval $\mathcal{I}_1$ the cross section obtained from the partial-wave analysis was used. The Mainz data cover the energy range 200 MeV to 800 MeV, and

Table 3.2. Contributions to the Baldin sum rule in units of $10^{-4}\,\mathrm{fm}^3$. The different energy intervals are defined in (3.45). For the low-energy interval $\mathcal{I}_1$, various solutions of the partial-wave analysis by Arndt et al. [34] were used to calculate their contribution. For $\mathcal{I}_3$, the result of [35] is given. The uncertainties include the experimental uncertainties of the data points, i.e. statistical and systematic uncertainties, and an estimated uncertainty for the integration method due to the spacing between the data points

SAID solution	$\mathcal{I}_1$ (SAID)	$\mathcal{I}_2$ (data)	$\mathcal{I}_3$ [35]
SM99K	1.39 ± 0.04	11.82 ± 0.36	0.73 ± 0.03
WI98K	1.19 ± 0.04	$-$	$-$
SP97K	1.19 ± 0.04	$-$	$-$
SM95	1.32 ± 0.04	$-$	$-$
Average	1.27 ± 0.10	11.82 ± 0.36	0.73 ± 0.03

at higher energies the Armstrong data complete the cross sections in the interval $\mathcal{I}_2$. Above 2000 MeV, the method of [35], which uses an overall function fitted to the cross section, was adopted. The results of the reevaluation [36] of the Baldin sum rule are tabulated in Table 3.2.[7] The low energy interval $\mathcal{I}_1$ contributes about 10% to the total sum rule and is strongly dependent on the partial-wave analysis used. Therefore, this contribution can only be estimated.

Using the average for $\mathcal{I}_1$, i.e. 1.3 ± 0.1, where the uncertainty is the standard deviation of the tabulated values, the Baldin sum rule can be evaluated as

$$\alpha + \beta = (13.8 \pm 0.4) \ . \tag{3.46}$$

This result is, of course, in agreement with that of [35], which is (13.69 ± 0.14), because the same cross section data were used. The difference reflects (i) the contribution of the energy region close to the π threshold, i.e. interval $\mathcal{I}_1$, and (ii) that in [35] the data points were fitted to obtain an excellent representative function for the integration. It customary to cite values of (14.2 ± 0.3) [20] or (14.2 ± 0.5) [21]. The first value was obtained from an analysis of the experimental absorption cross sections measured in the late 1960s at SLAC and DESY. The results of these experiments have been partially published by Caldwell et al. [37] and Ballam et al. [38]. The second value is based on the partial-wave analysis of Pfeil and Schwela [39] and Moorhouse et al. [40]. The new value of the Baldin sum rule (3.46), which is slightly lower than the older values, will be used throughout this book.

The fact that the photoabsorption cross section is well known allows, of course, the determination of additional moments of the cross section distribution, for example, the moment given in (3.43) corresponding to the quadrupole sum rule. As already mentioned, the influence of the threshold re-

---

[7] A trapezoidal-rule integration routine provided by CERNlib was used to evaluate the integral.

Table 3.3. Contributions to the quadrupole sum rule (3.43) from the different energy intervals defined in (3.45). The details of the calculation are the same as in Table 3.2, except that the interval $\mathcal{I}_3$ has been omitted owing to the strong convergence of the integral (Fig. 3.6)

SAID-Solution	$\mathcal{I}_1(SAID)$	$\mathcal{I}_2(data)$
SM99K	1.75 ± 0.05	4.26 ± 0.15
WI98K	1.48 ± 0.05	–
SP97K	1.48 ± 0.05	–
SM95	1.65 ± 0.05	–
Average	1.56 ± 0.13	4.26 ± 0.15

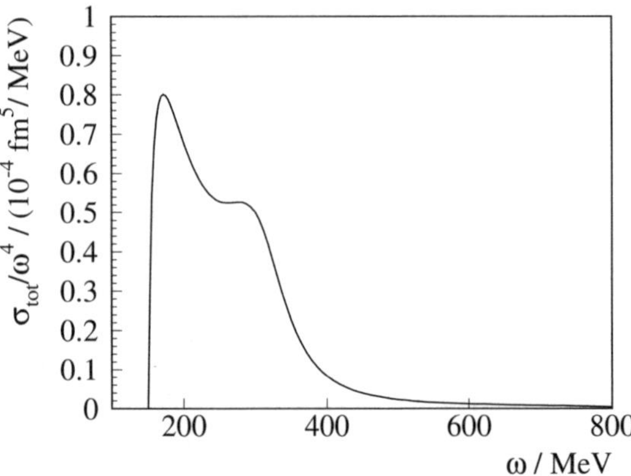

Fig. 3.6. Integrand of the quadrupole sum rule (3.43), as a function of the photon energy

gion increases here owing to the ω^{-4} weighting of the absorption cross section inside the integral. From Fig. 3.6, it can be seen that the integration converges very rapidly and the integral is already saturated at about 800 MeV. The contribution from the energy interval $\mathcal{I}_1$ (Table 3.3) is about 30% of the entire sum rule. The averages of the contributions and their maximum uncertainties are (1.56 ± 0.13) and (4.26 ± 0.15) for $\mathcal{I}_1$ and $\mathcal{I}_2$, respectively. From this, the quadrupole sum rule can be evaluated as

$$\alpha_2 + \beta_2 = 5.8 \pm 0.2 \,, \tag{3.47}$$

in the units given in Sect. A.1.

3.2 Dispersion Relations at Fixed t

At photon energies well below the π production threshold, i.e. at $\omega < 100$ MeV, the scattering cross sections are given by the proton's charge, mass and magnetic moment and two proton structure constants, the electric (α) and magnetic (β) polarizabilities, as discussed in Sect. 3.1.1. The low-energy expansion of the Compton scattering amplitude to the order $\omega\omega'$ can be used up to about 80 MeV incident photon energy to extract α and β from the measured differential cross sections. With increasing energy, the influence of higher-order terms has to be taken into account. The next-order terms are described by the four spin polarizabilities of the proton, γ_{E1}, γ_{M1}, γ_{E2} and γ_{M2}, two combinations of which are of special interest. These are the forward and backward spin polarizabilities $\gamma = -\gamma_{E1} - \gamma_{M1} - \gamma_{E2} - \gamma_{M2}$ and $\gamma_\pi = -\gamma_{E1} + \gamma_{M1} + \gamma_{E2} - \gamma_{M2}$, respectively. Since the cross sections are already dominated by the excitation of the Δ resonance at the energies under consideration, it is expected that the magnetic spin polarizabilities are determined mainly by the magnetic excitation of the Δ resonance.

At photon energies above 200 MeV, an expansion of the scattering amplitude in orders of the photon energy makes no sense and reliable model calculations do not exist. Therefore, a different approach has been chosen to investigate Compton scattering in the resonance regions, e.g. the dispersion relation approach of L'vov [41]. The advantage of this formalism is that the ingredients of the calculation are partially fixed by experimental results, i.e. the multipoles of π photoproduction, and the mechanisms which dominate Compton scattering can therefore be investigated in great detail.

Proofs of the fixed-t dispersion relations for elastic γp scattering and for the photoproduction of pions, based on the axioms of the relativistic quantum field theory of hadrons, have been given by Oehme and Taylor [42]. In [43], Oehme was able to show that the original methods of proof remain applicable "within the framework of Quantum Chromodynamics" not only for πN scattering but also for the above-mentioned reactions.

3.2.1 Invariant Scattering Amplitudes

The description of Compton scattering from a nucleon,

$$\gamma(k)\mathrm{N}(p) \rightarrow \gamma'(k')\mathrm{N}'(p') , \tag{3.48}$$

has been considered by Prange [44]. Assuming Lorentz, gauge and P–T invariance, the amplitude T_{fi} for elastic scattering is determined by the six invariant amplitudes T_i (see also Sect. 3.2.3):

$$
\begin{aligned}
T_{fi} = \bar{u}'(p')\epsilon'^{*\mu} &\left[-\frac{P'_\mu P'_\nu}{P'^2}\left(T_1 + \gamma \cdot KT_2\right) \right. \\
&- \frac{N_\mu N_\nu}{N^2}\left(T_3 + \gamma \cdot KT_4\right) + i\frac{P'_\mu N_\nu - P'_\nu N_\mu}{P'^2 K^2}\gamma_5 T_5 \\
&\left. + i\frac{P'_\mu N_\nu + P'_\nu N_\mu}{P'^2 K^2}\gamma_5\gamma \cdot KT_6 \right] \epsilon^\nu u(p) ,
\end{aligned}
\tag{3.49}
$$

where ϵ and ϵ' are the photon polarization vectors, u and u' are the nucleon bispinors normalized so that $\bar{u}u = 2m$, where m is the nucleon mass, and γ, γ_5 are the Dirac γ matrices. P', K, N and Q are orthogonal four-vectors defined by

$$P'_\mu = P_\mu - K_\mu \frac{P \cdot K}{K^2} \ , \quad P = \tfrac{1}{2}(p + p') \ , \quad K = \frac{1}{2}(k' + k) \ ,$$

$$N_\mu = \epsilon_{\mu\alpha\beta\gamma} P'^\alpha Q^\beta K^\gamma \ , \quad Q = \tfrac{1}{2}(p - p') = \frac{1}{2}(k' - k) \ , \tag{3.50}$$

where k, k' are the four-momenta of the photon and p, p' are those of the proton. In the laboratory and cm systems the differential cross sections for Compton scattering are evaluated according to the following equation:

$$\frac{\mathrm{d}\sigma}{\mathrm{d}\Omega} = \Phi^2 \, |T_{fi}|^2 \ , \quad \text{where } \Phi = \begin{cases} \frac{1}{8\pi m} \frac{\omega'}{\omega} & \text{(laboratory system)} \\[2mm] \frac{1}{8\pi\sqrt{s}} & \text{(cm system)} \end{cases} \ . \tag{3.51}$$

The definitions in (3.50) imply serious kinematical constraints. Because $K^2 \sim (1 - \cos\theta_\gamma)$, $P'^2 K^2 \sim (1 + \cos\theta_\gamma)$ and $N^2 \sim \sin\theta_\gamma{}^2$, the denominators in (3.49) vanish at forward and backward scattering angles. To formulate dispersion relations, it is convenient to introduce the following linear combinations of the amplitudes T_i [21, 41, 45]:

$$A_1 = \frac{1}{t}\left[T_1 + T_3 + \nu(T_2 + T_4)\right] \ , \tag{3.52}$$

$$A_2 = \frac{1}{t}\left[2T_5 + \nu(T_2 + T_4)\right] \ , \tag{3.53}$$

$$A_3 = \frac{m^2}{m^4 - su}\left[T_1 - T_3 - \frac{t}{4\nu}(T_2 - T_4)\right] \ , \tag{3.54}$$

$$A_4 = \frac{m^2}{m^4 - su}\left[2mT_6 - \frac{t}{4\nu}(T_2 - T_4)\right] \ , \tag{3.55}$$

$$A_5 = \frac{1}{4\nu}\left[T_2 + T_4\right] \ , \tag{3.56}$$

$$A_6 = \frac{1}{4\nu}\left[T_2 - T_4\right] \ . \tag{3.57}$$

The invariant amplitudes $A_i(\nu, t)$ are functions of the two variables ν and t, which are given by the Mandelstam variables:

$$\nu = \frac{s - u}{4m} = \omega + \frac{t}{4m} \ , \qquad s = (k + p)^2 \ , \tag{3.58}$$

$$u = (k - p')^2 \ , \qquad t = (k - k')^2 \ . \tag{3.59}$$

The form used by L'vov [41, 45] has been chosen for the invariant amplitudes A_i. This is not a commonly accepted description. The amplitudes introduced by Bardeen and Tung [46] are often referred to. In Sect. A.3 the relations between the different invariant amplitudes are summarized.

3.2.2 Dispersion Relations at Fixed t

By applying the basic formalism of dispersion relations, the real part of the invariant amplitudes A_i may be written at fixed t as [6, 41]

$$\operatorname{Re} A_i(\nu, t) = A_i^{\mathrm{Born}}(\nu, t) + A_i^{\mathrm{int}}(\nu, t) + A_i^{\mathrm{as}}(\nu, t) \,. \tag{3.60}$$

The first term on the right-hand side is the Born contribution. It is completely determined by the proton's charge, mass and magnetic moment:

$$A_i^{\mathrm{Born}}(\nu, t) = \frac{me^2 r_i(t)}{(s - m^2)(u - m^2)} \,, \tag{3.61}$$

with the definition[8] $e^2/4\pi = 1/137$ and where

$$r_1 = -2 + (\kappa^2 + 2\kappa)\frac{t}{4m^2} \,, \quad r_2 = 2\kappa + 2 + (\kappa^2 + 2\kappa)\frac{t}{4m^2} \,,$$
$$r_3 = \kappa^2 + 2\kappa \,, \qquad\qquad r_4 = \kappa^2 \,,$$
$$r_5 = \kappa^2 + 2\kappa \,, \qquad\qquad r_6 = -\kappa^2 - 2\kappa - 2 \,. \tag{3.62}$$

The second term on the right-hand side of (3.60), $A_i^{\mathrm{int}}(\nu, t)$, is the integral part, where the dispersion integral is evaluated from the π production threshold ν_{thr}, where $\omega_{\mathrm{thr}} = \nu_{\mathrm{thr}}(t) - t/4m = 150$ MeV, to a maximum energy ν_{max}, for which $\omega_{\mathrm{max}} = \nu_{\mathrm{max}}(t) - t/4m$ is taken to be equal to 1.5 GeV [41]:

$$A_i^{\mathrm{int}}(\nu, t) = \frac{2}{\pi}\mathcal{P} \int_{\nu_{\mathrm{thr}}}^{\nu_{\mathrm{max}}} \frac{\nu' \operatorname{Im} A_i(\nu', t)}{\nu'^2 - \nu^2} \mathrm{d}\nu' \,. \tag{3.63}$$

The third term in (3.60), $A_i^{\mathrm{as}}(\nu, t)$, is the asymptotic part, which contains all contributions to the dispersion integral above ν_{max}. The imaginary parts of the amplitudes A_i can be expressed mainly in terms of products of single-π photoproduction multipoles supplemented by two-pion contributions by making use of the unitarity condition. A detailed description of all the ingredients can be found in [41].

3.2.3 Constraints from Forward Scattering

To deduce some fundamental relations between the structure constants at low-energies, i.e. the polarizabilities, and the invariant amplitudes A_i, we turn to the helicity amplitudes for Compton scattering. The helicity of a particle, defined in the cm system, is given by the projection of its spin on the momentum axis. A photon can take the helicity states $\lambda_\gamma = \pm 1$ and a nucleon can have $\lambda_{\mathrm{N}} = \pm 1/2$. The net helicity of the photon–nucleon

---

[8] Compared with Sects. 3.1.1 and 3.1.2, the factor 4π is included here by virtue of the differential cross section defined in (3.51).

system, $\lambda = \lambda_\gamma - \lambda_N$, can therefore take the values $\pm 1/2$ and $\pm 3/2$. The transition from an initial state λ to a final state λ' is completely described by 16 helicity amplitudes $T_{\lambda'_\gamma \lambda'_N \lambda_\gamma \lambda_N}$. Owing to parity conservation and T invariance, these amplitudes reduced to six independent amplitudes [47]. In terms of the reduced helicity amplitudes τ_i, they read as [45]

$$T_{1\frac{1}{2}1\frac{1}{2}} = \cos\frac{\theta_\gamma}{2}\tau_1 \,, \qquad T_{-1\frac{1}{2}-1\frac{1}{2}} = \cos^3\frac{\theta_\gamma}{2}\tau_2 \,,$$

$$T_{1-\frac{1}{2}1\frac{1}{2}} = \cos^2\frac{\theta_\gamma}{2}\sin\frac{\theta_\gamma}{2}\tau_3 \,, \qquad T_{1\frac{1}{2}-1\frac{1}{2}} = \cos\frac{\theta_\gamma}{2}\sin^2\frac{\theta_\gamma}{2}\tau_4 \,,$$

$$T_{-1-\frac{1}{2}1\frac{1}{2}} = \sin\frac{\theta_\gamma}{2}\tau_5 \,, \qquad T_{1-\frac{1}{2}-1\frac{1}{2}} = \sin^3\frac{\theta_\gamma}{2}\tau_6 \,. \tag{3.64}$$

It becomes apparent that for forward scattering, where $\theta_\gamma = 0$ and therefore $\sin\theta_\gamma/2 = 0$, only those amplitudes related to τ_1 and τ_2 survive, which correspond to the helicity states $|\lambda| = 1/2$ and $|\lambda| = 3/2$. The optical theorem reads, in the normalization defined in (3.51), as

$$\mathrm{Im}\,\tau_{1,2} = 2m\nu\sigma_{1/2,3/2} \,, \tag{3.65}$$

$$\Rightarrow \mathrm{Im}(\tau_1 + \tau_2) = 2m\nu(\sigma_{1/2} + \sigma_{3/2}) = 4m\nu\sigma_{\mathrm{tot}} \,, \tag{3.66}$$

where for $t = 0$, $\nu = \omega$ is the laboratory photon energy. From (A.18) and (A.21) it follows for $t = 0$,

$$A_3(\nu, t = 0) = \frac{1}{(s - m^2)^3}\left[m^3(\tau_1 + \tau_2) - 2m^2\sqrt{s}\tau_3\right] \,, \tag{3.67}$$

$$A_6(\nu, t = 0) = \frac{1}{(s - m^2)^3}\left[-\frac{m}{2}(s + m^2)(\tau_1 + \tau_2) + 2m^2\sqrt{s}\tau_3\right] \,, \tag{3.68}$$

and therefore

$$A_3(\nu, 0) + A_6(\nu, 0) = -\frac{1}{8m\nu^2}(\tau_1 + \tau_2) \,. \tag{3.69}$$

Considering the imaginary parts and inserting (3.66), we obtain:

$$\mathrm{Im}\,(A_3(\nu, 0) + A_6(\nu, 0)) = \mathrm{Im}\,A_{3+6}(\nu, 0) = -\frac{1}{2\nu}\sigma_{\mathrm{tot}} \,. \tag{3.70}$$

With this input, the Baldin sum rule (3.41) can be evaluated from the imaginary parts of the amplitudes A_3 and A_6 at $t = 0$:

$$\alpha + \beta = -\frac{1}{\pi^2}\int_{\nu_{\mathrm{thr}}}^{\infty}\frac{\mathrm{Im}\,A_{3+6}(\nu', 0)}{\nu'}\,\mathrm{d}\nu' \,. \tag{3.71}$$

A simplification may be achieved if the non-Born (nB) part in (3.60), $A_i^{\mathrm{nB}} = A_i^{\mathrm{int}} + A_i^{\mathrm{as}}$, for the sum of the amplitudes $A_3 + A_6$ at $\nu = t = 0$, is written as

$$A^{\mathrm{nB}}_{3+6}(0,0) = A^{\mathrm{int}}_{3+6}(0,0) + A^{\mathrm{as}}_{3+6}(0,0)$$ (3.72)

$$= \frac{2}{\pi} \int\limits_{\nu_{\mathrm{thr}}}^{\infty} \frac{\mathrm{Im}\, A_{3+6}(\nu',0)}{\nu'}\, \mathrm{d}\nu' \ .$$ (3.73)

It follows that the Baldin sum rule fixes the non-Born part of the amplitudes A_3 and A_6 in the forward direction at $\nu = t = 0$:

$$\alpha + \beta = -\frac{1}{2\pi} A^{\mathrm{nB}}_{3+6}(0,0) = -\frac{1}{2\pi}\left[A^{\mathrm{int}}_{3+6}(0,0) + A^{\mathrm{as}}_{3+6}(0,0)\right] \ .$$ (3.74)

The same formalism may be applied to the difference $\tau_1 - \tau_2$. From (A.19) and (3.65), it follows that for $t = 0$,

$$\mathrm{Im}\, A_4(\nu, t = 0) = \frac{m^3}{(s-m^2)^3}\mathrm{Im}(\tau_1 - \tau_2)$$

$$= \frac{m}{4\nu^2}(\sigma_{1/2} - \sigma_{3/2}) \ .$$ (3.75)

Rewriting the non-Born part of A_4 at $\nu = t = 0$ and inserting (3.75), we immediately obtain

$$A^{\mathrm{nB}}_4(0,0) = \frac{m}{2\pi} \int\limits_{\nu_{\mathrm{thr}}}^{\infty} \frac{\sigma_{1/2} - \sigma_{3/2}}{\nu'^3}\, \mathrm{d}\nu' \ ,$$ (3.76)

which is equivalent to the forward spin polarizability sum rule given in (3.44). This is another constraint on the invariant amplitudes as far as A_4 is concerned:

$$\gamma = \frac{1}{2\pi m} A^{\mathrm{nB}}_4(0,0) = \frac{1}{2\pi m}\left[A^{\mathrm{int}}_4(0,0) + A^{\mathrm{as}}_4(0,0)\right] \ .$$ (3.77)

It should be mentioned that the sum rules above can also be derived from an expansion of the forward scattering amplitude, which reads as follows in terms of the invariant amplitudes (laboratory system) [15]:

$$\frac{1}{8\pi m}T_{fi}\bigg|_{\theta=0^\circ} = \frac{\omega^2}{2\pi}\left[-\boldsymbol{\epsilon}' \cdot \boldsymbol{\epsilon}\,(A_3 + A_6) + i\boldsymbol{\sigma} \cdot (\boldsymbol{\epsilon}' \times \boldsymbol{\epsilon})\frac{\omega}{m}A_4\right] \ .$$ (3.78)

After separating the Born contribution by writing $A_i = A^{\mathrm{B}}_i + A^{\mathrm{nB}}_i$, the non-Born parts, which are functions of the variable ν^2, may be expanded in a power series in this variable:

$$A^{\mathrm{nB}}_i(\nu,0) = A^{\mathrm{nB}}_i(0,0) + \nu^2 a_{i,\nu} + \cdots \ .$$ (3.79)

The abbreviation

$$a_{i,\nu} = \left(\frac{\partial A^{\mathrm{nB}}_i}{\partial \nu^2}\right)_{\nu=t=0}$$ (3.80)

represents the derivatives with respect to ν^2 of the non-Born parts of the amplitudes A_i at zero energy and zero momentum transfer. The forward scattering amplitude (3.78) can be recast with the help of (3.61), and we obtain the following low-energy expansion [15]:

$$\frac{1}{8\pi m}T_{fi}\bigg|_{\theta=0^\circ} = \left[-\frac{e^2}{m} - \frac{\omega^2}{2\pi}\left(A_3^{\mathrm{nB}}(0,0) + A_6^{\mathrm{nB}}(0,0)\right)\right.$$
$$\left. -\frac{\omega^4}{2\pi}\left(a_{3,\nu} + a_{6,\nu}\right)\right]\boldsymbol{\epsilon}'\cdot\boldsymbol{\epsilon}$$
$$+\mathrm{i}\omega\left[-\frac{e^2\kappa^2}{2m^2} + \frac{\omega^2}{2\pi m}A_4^{\mathrm{nB}}(0,0)\right]\boldsymbol{\sigma}\cdot(\boldsymbol{\epsilon}'\times\boldsymbol{\epsilon}) . \qquad (3.81)$$

The sum rules above are immediately obtained by a comparison with the low-energy expansion given in (3.27). In addition, the quadrupole sum rule (3.43) can be written in terms of the derivatives of the amplitudes A_3 and A_6 at $\nu = t = 0$:

$$\alpha_2 + \beta_2 = -\frac{1}{2\pi}\left(a_{3,\nu} + a_{6,\nu}\right) . \qquad (3.82)$$

3.2.4 Constraints from Backward Scattering

In continuation of the last section, we now derive constraints on the invariant amplitudes from backward scattering, i.e. $\theta_\gamma = 180^\circ$. The backward scattering amplitude in terms of the A_i (laboratory system) [15] is given by

$$\frac{1}{8\pi m}T_{fi}\bigg|_{\theta=180^\circ} = -\frac{\omega\omega'}{2\pi}\left[N(t)\left(A_1 - \frac{t}{4m^2}A_5\right)\boldsymbol{\epsilon}'\cdot\boldsymbol{\epsilon}\right.$$
$$\left. +\mathrm{i}\frac{\nu}{N(t)m}\left(A_2 + N^2(t)A_5\right)\boldsymbol{\sigma}\cdot(\boldsymbol{\epsilon}'\times\boldsymbol{\epsilon})\right] , \qquad (3.83)$$

where $N(t) = \sqrt{1 - t/4m^2}$. Since the invariant amplitudes depend on $\nu^2 = \omega\omega' + t^2/16m^2$ and $t = -2\omega\omega'(1 - z)$ at backward angles, one has to expand the non-Born parts in powers of $\omega\omega'$ and t:

$$A_i^{\mathrm{nB}}(\nu, t) = A_i^{\mathrm{nB}}(0,0) + \omega\omega' a_{i,\nu} + t a_{i,t} \qquad (3.84)$$
$$= A_i^{\mathrm{nB}}(0,0) + \omega\omega'\left(a_{i,\nu} - 2(1 - z)a_{i,t}\right) . \qquad (3.85)$$

The derivatives $a_{i,\nu}$ have been defined in (3.80), and

$$a_{i,t} = \left(\frac{\partial A_i^{\mathrm{nB}}}{\partial t}\right)_{\nu=t=0} . \qquad (3.86)$$

From the above equations we finally obtain the low-energy expansion of the scattering amplitude at $\theta_\gamma = 180^\circ$:

$$\frac{1}{8\pi m}T_{fi}\bigg|_{\theta=180°} = \bigg[-\frac{e^2}{m} - \frac{\omega\omega'}{2\pi}A_1^{\text{nB}}(0,0)$$

$$-\frac{(\omega\omega')^2}{2\pi}\left(a_{1,\nu} - 4a_{1,t} + \frac{A_5^{\text{nB}}(0,0)}{m^2}\right)\bigg]N(t)\boldsymbol{\epsilon}'\cdot\boldsymbol{\epsilon}$$

$$+i\sqrt{\omega\omega'}\bigg[\frac{e^2}{m}\frac{\kappa^2 + 4\kappa + 2}{2m}$$

$$-\frac{\omega\omega'}{2\pi m}\left(A_2^{\text{nB}}(0,0) + A_5^{\text{nB}}(0,0)\right)\bigg]\boldsymbol{\sigma}\cdot(\boldsymbol{\epsilon}'\times\boldsymbol{\epsilon}) \ . \quad (3.87)$$

Comparing (3.87) with the standard low-energy expansion of the backward scattering amplitude [15],

$$\frac{1}{8\pi m}T_{fi}\bigg|_{\theta=180°} = \bigg[-\frac{e^2}{m} + \omega\omega'(\alpha - \beta) + (\omega\omega')^2(\alpha_2 - \beta_2)\bigg]N(t)\boldsymbol{\epsilon}'\cdot\boldsymbol{\epsilon}$$

$$+i\sqrt{\omega\omega'}\bigg[\frac{e^2}{m}\frac{\kappa^2 + 4\kappa + 2}{2m} + \omega\omega'\gamma_\pi\bigg]\boldsymbol{\sigma}\cdot(\boldsymbol{\epsilon}'\times\boldsymbol{\epsilon}) \ , (3.88)$$

we obtain the following constraints on the invariant amplitudes at $\nu = t = 0$:

$$\alpha - \beta = -\frac{1}{2\pi}A_1^{\text{nB}}(0,0) \ , \qquad\qquad (3.89)$$

$$\alpha_2 - \beta_2 = -\frac{1}{2\pi}\bigg[a_{1,\nu} - 4a_{1,t} + \frac{A_5^{\text{nB}}(0,0)}{m^2}\bigg] \ , \qquad (3.90)$$

$$\gamma_\pi = -\frac{1}{2\pi m}\big[A_2^{\text{nB}}(0,0) + A_5^{\text{nB}}(0,0)\big] \ . \qquad (3.91)$$

In accordance with (3.77) the quantity γ_π in (3.91) is called the backward spin polarizability. This gives additional access to the polarizabilities for describing the electromagnetic structure of the proton at low photon energies.

3.2.5 Asymptotic Contributions

The relations which constrain the invariant amplitudes A_i are summarized in Table 3.4 . For forward scattering angles and low photon energies, the spin-independent part of the scattering amplitude is determined by the non-Born parts of the amplitudes A_3 and A_6 at $\nu = t = 0$. At higher orders in ω, it continues as given by the derivatives with respect to ν^2 of these amplitudes. The reader should be aware that in the laboratory system, ν is given by

$$\nu = \frac{1}{2}(\omega + \omega') \ , \qquad\qquad (3.92)$$

where ω' is the energy of the scattered photon. Therefore, at $\theta_\gamma = 0$, ν coincides with the incident photon energy ω. At backward angles, the amplitude A_1 fixes the low-energy behavior of the spin-independent part and the higher

Table 3.4. Constraints on the invariant amplitudes A_i arising from forward and backward scattering

---

Forward scattering

$$\alpha + \beta = -\frac{1}{2\pi} \left[A_{3+6}^{\text{int}}(0,0) + A_{3+6}^{\text{as}}(0,0) \right]$$

$$\alpha_2 + \beta_2 = -\frac{1}{2\pi} \left[a_{3,\nu} + a_{6,\nu} \right]$$

$$\gamma = \frac{1}{2\pi m} \left[A_4^{\text{int}}(0,0) + A_4^{\text{as}}(0,0) \right]$$

Backward scattering

$$\alpha - \beta = -\frac{1}{2\pi} \left[A_1^{\text{int}}(0,0) + A_1^{\text{as}}(0,0) \right]$$

$$\alpha_2 - \beta_2 = -\frac{1}{2\pi} \left[a_{1,\nu} - 4a_{1,t} + \frac{1}{m^2} \left(A_5^{\text{int}}(0,0) + A_5^{\text{as}}(0,0) \right) \right]$$

$$\gamma_\pi = -\frac{1}{2\pi m} \left[A_{2+5}^{\text{int}}(0,0) + A_{2+5}^{\text{as}}(0,0) \right]$$

---

orders are determined by the derivatives with respect to ν^2 and t. Additionally, the amplitude A_5 appears.

The amplitudes $A_i(\nu, t)$ are convergent in the high-energy limit at fixed t [41]. For $\nu \to \infty$, it follows that $A_i \to 0$, except for $A_1(\nu, t)$ and $A_2(\nu, t)$, which are assumed to be proportional to $\nu^{\alpha(t)}$, where $\alpha(t) \leq 1$ is the Regge trajectory. Therefore, the latter two amplitudes require special consideration. In the case of the amplitude $A_1(\nu, t)$, the non-Born part taken at $\nu = t = 0$ is related to the difference $\alpha - \beta$ between the electromagnetic polarizabilities, whereas the amplitude $A_2(\nu, t)$ is related to the backward spin polarizability γ_π (Table 3.4). Note that neither of these amplitudes contributes to forward scattering. The integral parts are completely determined by the multipoles of π photoproduction. From the decomposition of the amplitudes T_i with respect to A_i (A.23), it follows that only T_1 and T_3 are related to A_1, and A_2 appears only in T_5. In the scattering amplitude in (3.49), T_1, T_3 and T_5 appear in scalar (~ 1) and pseudoscalar ($\sim \gamma^5$) terms, respectively. From the decomposition of the A_i into the helicity amplitudes (A.16), one can see that for $\theta_\gamma = 180°$ the amplitudes A_1 and A_2 are related to a photon and proton helicity flip transition. Since only the directions of the momenta change at this extreme angle, no spin flip occurs, i.e. the spin alignment of the photon and the proton is the same before and after the scattering process. This favours the exchange of spinless particles. One may therefore conclude that the asymptotic contributions are related to the t-channel exchange of the lightest scalar and pseudoscalar particles which are the σ and π^0 meson.

Such t-channel exchanges are commonly described by a pole diagram located at the meson mass, which in the case of a π^0 exchange is called the

Low amplitude [41]

$$A_2^{\text{as}}(\nu, t) \approx A_2^{\pi^0}(t) = \frac{g_{\pi\text{NN}} F_{\pi^0\gamma\gamma}}{t - m_{\pi^0}^2} \tau_3 F_\pi(t)\,,\qquad (3.93)$$

where τ_3 is the isospin factor, equal to $+1$ and -1 for the proton and neutron, respectively. The πNN and $\pi^0\gamma\gamma$ coupling constants have been discussed in Sect. 3.1.1 and are given in (3.13). The Low amplitude can be extended by an off-shell form factor $F_\pi(t) = (\Lambda_\pi^2 - m_\pi^2)/(\Lambda_\pi^2 - t)$ with a cutoff parameter $\Lambda_\pi = 700$ MeV. The same procedure may be applied to the σ exchange in the t channel:

$$A_1^{\text{as}}(\nu, t) \approx \frac{g_{\sigma\text{NN}} F_{\sigma\gamma\gamma}}{t - m_\sigma^2}\,.\qquad (3.94)$$

The absence of a form factor is motivated by the unknown mass of the σ meson, which serves as a parameter to reproduce the cross sections at backward angles. Up to now, the σ meson has never been directly observed, but evidence has been found in the analysis of $\pi\pi$ scattering [48, 49]. One may associate a correlated $\pi\pi$ pair with this particle or identify it with the f_0 meson [13].

3.2.6 Higher-Order Electromagnetic Polarizabilities

As already mentioned in Sect. 3.1.3, the higher-order polarizabilities α_2, β_2 are linear combinations of the dispersion polarizabilities $\alpha_{\text{E}\nu}$, $\beta_{\text{M}\nu}$ and the quadrupole polarizabilities $\alpha_{\text{E}2}$, $\beta_{\text{M}2}$:

$$\alpha_2 + \beta_2 = \alpha_{\text{E}\nu} + \beta_{\text{M}\nu} + \frac{1}{12}\alpha_{\text{E}2} + \frac{1}{12}\beta_{\text{M}2}\,,\qquad (3.95)$$

$$\alpha_2 - \beta_2 = \alpha_{\text{E}\nu} - \beta_{\text{M}\nu} - \left(\frac{1}{12}\alpha_{\text{E}2} - \frac{1}{12}\beta_{\text{M}2}\right)\,.\qquad (3.96)$$

Without restrictions to kinematical extremes, such as forward or backward scattering, a direct comparison between the low-energy expansion of the scattering amplitudes to the order $\mathcal{O}((\omega\omega')^2)$ and the low-energy expansion of the invariant amplitudes $A_i(\nu, t)$ gives a relation between the different contributions and the invariant amplitudes [15]. The dispersion polarizabilities read as

$$\alpha_{\text{E}\nu} = -\frac{1}{4\pi}\left(a_{3,\nu} + a_{6,\nu} + a_{1,\nu} + \frac{A_5^{\text{nB}}(0,0)}{m^2}\right.$$
$$\left. -a_{3,t} - a_{6,t} - 3a_{1,t} + \frac{A_3^{\text{nB}}(0,0)}{4m^2}\right)\,,\qquad (3.97)$$

$$\beta_{\text{M}\nu} = -\frac{1}{4\pi}\left(a_{3,\nu} + a_{6,\nu} - a_{1,\nu} - \frac{A_5^{\text{nB}}(0,0)}{m^2}\right.$$
$$\left. -a_{3,t} - a_{6,t} + 3a_{1,t} + \frac{A_3^{\text{nB}}(0,0)}{4m^2}\right)\,.\qquad (3.98)$$

The quantities $a_{i,\nu}$ and $a_{i,t}$ have been defined in (3.80) and (3.86), respectively. For the quadrupole polarizabilities we obtain

$$\alpha_{\mathrm{E2}} = -\frac{3}{\pi}\left(a_{3,t} + a_{6,t} + a_{1,t} - \frac{A_3^{\mathrm{nB}}(0,0)}{4m^2}\right), \tag{3.99}$$

$$\beta_{\mathrm{M2}} = -\frac{3}{\pi}\left(a_{3,t} + a_{6,t} - a_{1,t} - \frac{A_3^{\mathrm{nB}}(0,0)}{4m^2}\right). \tag{3.100}$$

In principle, these relations can be used to evaluate all the electromagnetic polarizabilities of higher order. Using the π photoproduction multipoles of Arndt et al. [34], solution SM99K, the values of the higher-order electromagnetic polarizabilities tabulated in Table 3.1 have been calculated. Owing to the constraints from forward and backward scattering (Table 3.4), some linear combinations are strongly restricted by the experimental data. For example, the quadrupole sum rule has been evaluated as $\alpha_2 + \beta_2 = 5.8 \pm 0.2$ (3.47).

3.2.7 Spin Polarizabilities

The forward and backward spin polarizabilities γ and γ_π enter into the differential cross section at these extreme scattering angles. These spin polarizabilities are linear combinations of the four common spin polarizabilities γ_1 to γ_4 introduced by Ragusa [16] in the low-energy expansion of the scattering amplitude.

In the notation of Babusci et al. [15], the four spin polarizabilities ((3.20) and (3.21)) are related to the non-Born parts of the invariant scattering amplitudes at $\nu = t = 0$ as follows:

$$\gamma_{\mathrm{E1}} = \frac{1}{8\pi m}\left(A_6^{\mathrm{nB}}(0,0) - A_4^{\mathrm{nB}}(0,0) + 2A_5^{\mathrm{nB}}(0,0) + A_2^{\mathrm{nB}}(0,0)\right), \tag{3.101}$$

$$\gamma_{\mathrm{M1}} = \frac{1}{8\pi m}\left(A_6^{\mathrm{nB}}(0,0) - A_4^{\mathrm{nB}}(0,0) - 2A_5^{\mathrm{nB}}(0,0) - A_2^{\mathrm{nB}}(0,0)\right), \tag{3.102}$$

$$\gamma_{\mathrm{E2}} = -\frac{1}{8\pi m}\left(A_6^{\mathrm{nB}}(0,0) + A_4^{\mathrm{nB}}(0,0) + A_2^{\mathrm{nB}}(0,0)\right), \tag{3.103}$$

$$\gamma_{\mathrm{M2}} = -\frac{1}{8\pi m}\left(A_6^{\mathrm{nB}}(0,0) + A_4^{\mathrm{nB}}(0,0) - A_2^{\mathrm{nB}}(0,0)\right). \tag{3.104}$$

From these equations, γ and γ_π can be obtained via

$$\gamma = -\gamma_{\mathrm{E1}} - \gamma_{\mathrm{M1}} - \gamma_{\mathrm{E2}} - \gamma_{\mathrm{M2}}, \tag{3.105}$$

$$\gamma_\pi = -\gamma_{\mathrm{E1}} + \gamma_{\mathrm{M1}} + \gamma_{\mathrm{E2}} - \gamma_{\mathrm{M2}}, \tag{3.106}$$

from which the constraints in Table 3.4 follow. Using the π photoproduction multipoles of Arndt et al. [34], solution SM99K, the values of the spin polarizabilities tabulated in Table 3.1 were calculated.

References

1. F. E. Low, Phys. Rev. **96** (1954) 1428
2. M. Gell-Mann, M. L. Goldberger, W. Thirring, Phys. Rev. **95** (1954) 1612
3. M. Gell-Mann, M. L. Goldberger, Phys. Rev. **96** (1954) 1433
4. A. Klein, Phys. Rev. **99** (1955) 998
5. V. A. Petrun'kin, Sov. Phys. JETP **13** (1961) 808
6. V. A. Petrun'kin, Sov. J. Part. Nucl. **12** (1981) 278
7. J. L. Powell, Phys. Rev. **75** (1949) 32
8. O. Klein, Y. Nishina, Z. Phys. **52** (1929) 853
9. P. S. Baranov, L. V. Fil'kov, L. N. Shtarkov, JETP Lett. **20** (1974) 353
10. I. Guiaşu, E. E. Radescu, Phys. Rev. D **18** (1978) 651
11. I. Guiaşu, C. Pomponiu, E. E. Radescu, Ann. Phys. **114** (1978) 296
12. R. A. Arndt, R. L. Workman, M. M. Pavan, Phys. Rev. C **49** (1994) 2729
13. K. Hagiwara et al., Phys. Rev. D **66** (2002) 010001, *Review of Particle Physics*, Particle Data Group, http://pdg.lbl.gov
14. M. V. Terent'ev, Sov. J. Nucl. Phys. **16** (1973) 576
15. D. Babusci et al., Phys. Rev. C **58** (1998) 1013
16. S. Ragusa, Phys. Rev. D **47** (1993) 3757
17. F. J. Federspiel et al., Phys. Rev. Lett. **67** (1991) 1511
18. B. E. MacGibbon et al., Phys. Rev. C **52** (1995) 2097
19. A. Zieger et al., Phys. Lett. B **278** (1992) 34
20. M. Damashek, F. J. Gilman, Phys. Rev. D **1** (1970) 1319
21. A. I. L'vov, V. A. Petrun'kin, S. A. Startsev, Sov. J. Nucl. Phys. **29** (1979) 651
22. B.H. Bransden, R.G. Moorhouse, *The Pion–Nucleon System*, Princeton University Press, Princton, NJ, 1973
23. B.H. Bransden, D. Evans, J.V. Major, *The Fundamental Particles*, Van Nostrand Reinhold, London, 1973
24. R. Kronig, J. Opt. Soc. Am. **12** (1926) 547
25. H. A. Kramers, Atti Congr. Intern. Fisici Como (1927)
26. D. Drechsel, B. Pasquini, M. Vanderhaeghen, Phys. Rep. **378** (2003) 99. [This comprehensive article was published after the completion of the manuscript of the present book]
27. R. L. Walker, Phys. Rev. **182** (1969) 1729
28. H. Rollnik, P. Stichel, *Compton Scattering*, STMP, Vol. 79, Springer, Berlin, Heidelberg, 1976, p. 1
29. J. D. Bjorken, S. D. Drell, *Relativistic Quantum Fields*, McGraw-Hill, New York, 1965
30. A. M. Baldin, Nucl. Phys. **18** (1960) 310
31. S. B. Gerasimov, Sov. J. Nucl. Phys. **2** (1966) 430; S. D. Drell, A. C. Hearn, Phys. Rev. Lett. **16** (1966) 908
32. T. A. Armstrong et al., Phys. Rev. D **5** (1972) 1640
33. M. MacCormick et al., Phys. Rev. C **53** (1996) 41
34. R. A. Arndt et al., Phys. Rev. C **53** (1996) 430; the SAID database can be accessed via http://gwdac.phys.gwu.edu
35. D. Babusci, G. Giordano, G. Matone, Phys. Rev. C **57** (1998) 291
36. V. Olmos de León et al., Eur. Phys. J. A **10** (2001) 207
37. D. O. Caldwell et al., Phys. Rev. Lett. **23** (1969) 1256

38. J. Ballam et al., Phys. Rev. Lett. **21** (1968) 1544; J. Ballam et al., Phys. Rev. Lett. **23** (1969) 498
39. W. Pfeil, D. Schwela, Nucl. Phys. B **45** (1972) 379
40. R. G. Moorhouse, H. Oberlack, A. H. Rosenfeld, Phys. Rev. D **9** (1974) 1
41. A. I. L'vov, V. A. Petrun'kin, M. Schumacher, Phys. Rev. C **55** (1997) 359
42. R. Oehme, J.G. Taylor, Phys. Rev. **113** (1959) 371
43. R. Oehme, πN Newsletter **7** (1992) 1; J. Mod. Phys. Lett. A **8** (1993) 1533
44. R. E. Prange, Phys. Rev. **100** (1958) 240
45. A. I. L'vov, Sov. J. Nucl. Phys. **34** (1981) 597
46. W. A. Bardeen, W. K. Tung, Phys. Rev. **173** (1968) 1423
47. A. C. Hearn, E. Leader, Phys. Rev. **126** (1962) 789
48. G. Colangelo, J. Gasser, H. Leutwyler, Nucl. Phys. B **603** (2001) 125
49. M. Ishida, Prog. Theor. Phys. Suppl. 149 (2003) 190, *Proc. of YITP-RCNP Workshop on Chiral Restoration in Nuclear Medium*, Kyoto, Japan, 7–9 October 2002

4 Experiments on the Proton

4.1 Experiment at Low Photon Energies

4.1.1 TAPS Experiment

Compton scattering from the proton in the energy range 55 MeV to 165 MeV [1, 2] has been investigated with the TAPS detector system (Sect. A.7), set up at the photon beam at MAMI (Sect. A.4). The energy of the incident electron beam was chosen to be 180 MeV. The emitted bremsstrahlung beam was collimated, resulting in a tagging efficiency of about 17% (measured at low intensity with a bismuth germanate (BGO) detector in the direct photon beam). This low tagging efficiency results from the low incident electron energy and the collimation system behind the tagger.

The target consisted of a Kapton cylinder 20 cm long filled with liquid hydrogen. The target thickness was $N_{\mathrm{T}} = (8.66 \pm 0.18) \times 10^{23}$ cm^{-2}. Data obtained from 200 h of beam time were analyzed. The basic features of the experiment are summarized in Table 4.1.

The scattered photons were detected with the six blocks (A to F) of the TAPS detector only. The forward wall was not included in the setup during this experiment. Since the recoiling protons could not be detected, a single-particle trigger had to be used, i.e. the minimum block multiplicity was set to 1 in order to create the trigger signal. Therefore, this minimum-bias trigger included all kinds of background events. There were cosmic-ray events, which had not been suppressed by any active shield. There was electromagnetic background from the beam collimation system and from the target itself, which hit the blocks installed close to the beam. These sources of background could be partially suppressed by restrictive time cuts.

The "missing energy" of the photon ΔE_γ is defined here as the difference between the measured incident photon energy (measured by the tagger) and the expected incident photon energy calculated from the measured momentum of the scattered photon assuming Compton kinematics (Sect. A.5). It permits one to combine several tagging channels in one spectrum without losing the energy resolution provided by the tagger. Such spectra are shown in Fig. 4.1 for $\theta_\gamma = 59°$ and $133°$ at incident photon energies of 89.1 MeV and 157.3 MeV. The peak of elastically scattered photons appears at zero missing energy. At forward angles and low photon energies, the background

Table 4.1. Overview of the TAPS experiment

θ_γ / deg.	E_γ / MeV	ε_{tag} / %	Beam time / h
59–155	55–165	17	200

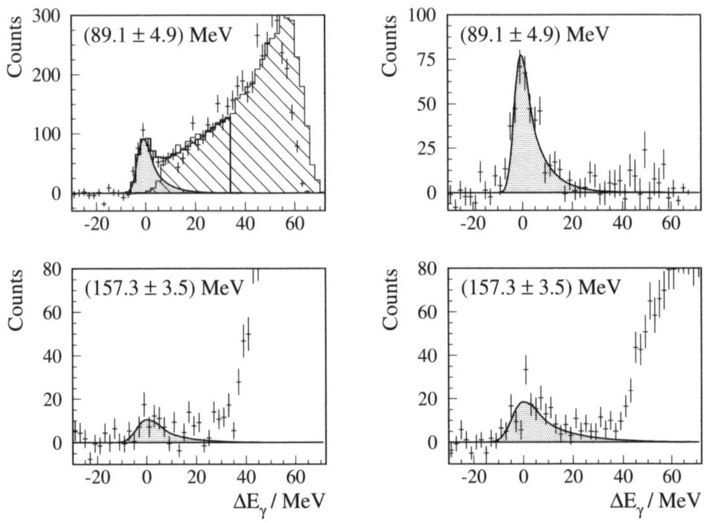

Fig. 4.1. Missing-energy spectra (*crosses*) at $\theta_\gamma = 59°$ (*left*) and $\theta_\gamma = 133°$ (*right*). The *dotted areas* correspond to the simulated response to elastically scattered photons. *Top*: spectrum below π production threshold. The *hatched area* is the measured missing-energy spectrum of charged particles adjusted to the photon missing-energy spectrum. *Bottom*: spectrum above π production threshold. The rise in intensity at larger missing energy is due to π^0 photoproduction

produced in the target originates mainly from pair production. Such events could be only partially suppressed by the veto detectors. By identifying these charged particles with the veto detectors, the missing-energy distributions of these events could be measured and then adjusted to the measured photon missing-energy spectra (hatched area in Fig. 4.1). Above the π production threshold, additional photons arise owing to π^0 photoproduction followed by immediate 2γ decay. These events are clearly separated from the elastically scattered photons. Thus, the number of scattered photons could be extracted using the simulated response of the TAPS detector to scattered photons.

4.1.2 Determination of the Electromagnetic Polarizabilities

The results of the experiment in comparison with previous data are shown in Fig. 4.2. The systematic (normalization) errors of $\pm 3\%$, which affect all measured cross sections in common, arise from uncertainties in the photon

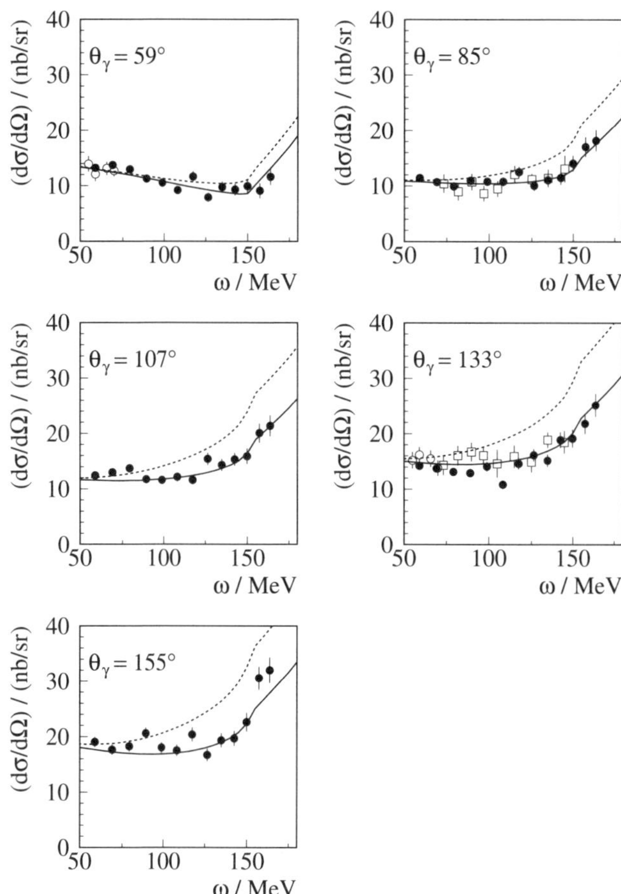

Fig. 4.2. The measured differential cross sections in the laboratory system as obtained by the TAPS experiment ($\bullet$) [1, 2]. Additional data have been taken from [3] ($\circ$) and [4] ($\square$). The *solid line* shows the result of a calculation performed with the dispersion relation approach described in Sect. 3.2 using the SAID SM99K π-photoproduction multipoles of Arndt et al. [5]. The polarizabilities were chosen to be $\alpha_{\mathrm{p}} + \beta_{\mathrm{p}} = 13.8$, $\alpha_{\mathrm{p}} - \beta_{\mathrm{p}} = 10.6$ and $\gamma_{\pi}^{(\mathrm{p})} = -37.1$. The *dashed line* shows the result of a similar calculation using $g_{\pi\mathrm{NN}}F_{\pi^0\gamma\gamma} > 0$

flux ($\pm 2\%$) and target density ($\pm 2\%$) combined in quadrature. The effective solid angles were determined by Monte Carlo simulations. Errors from uncertainties in the geometry of the experiment and from the statistics of the simulation may be treated as individual (random) errors ("random systematic" errors). They are estimated to be $\pm 5\%$. During the fitting procedure, as explained below, this error has therefore been added in quadrature to the statistical errors of the individual data points.

Table 4.2. The polarizabilities $\alpha_{\rm p}$ and $\beta_{\rm p}$ as obtained by fitting the differential cross sections obtained from different experiments. The uncertainties given are the statistical and systematic ones. "fixed" denotes that $\alpha_{\rm p} + \beta_{\rm p} = 13.8 \pm 0.4$ was included as a single data point. The data of Zieger et al. have been fitted with $\alpha_{\rm p} - \beta_{\rm p}$ as the only parameter. The last row shows the result of the global fit, for which all data sets were allowed to vary in accordance with their systematic uncertainties

Data		$\alpha_{\rm p} + \beta_{\rm p}$ fixed	$\alpha_{\rm p} + \beta_{\rm p}$ free
TAPS	$\alpha_{\rm p}$	$12.1 \pm 0.4 \mp 1.0$	$11.9 \pm 0.5 \mp 1.3$
(this work)	$\beta_{\rm p}$	$1.6 \pm 0.4 \pm 0.8$	$1.2 \pm 0.7 \pm 0.3$
MacGibbon	$\alpha_{\rm p}$	$11.9 \pm 0.5 \mp 0.8$	$12.6 \pm 1.2 \mp 1.3$
[4]	$\beta_{\rm p}$	$1.9 \pm 0.5 \pm 0.8$	$3.0 \pm 1.8 \pm 0.1$
Federspiel	$\alpha_{\rm p}$	$10.8 \pm 2.2 \mp 1.3$	$10.1 \pm 2.6 \mp 2.0$
[3]	$\beta_{\rm p}$	$3.0 \pm 2.2 \pm 1.3$	$2.0 \pm 3.3 \pm 0.3$
Zieger	$\alpha_{\rm p} - \beta_{\rm p}$	$6.4 \pm 2.3 \pm 1.9$	
[6]			
Global	$\alpha_{\rm p}$	$12.1 \pm 0.3 \mp 0.4$	$11.9 \pm 0.5 \mp 0.5$
fit	$\beta_{\rm p}$	$1.6 \pm 0.4 \pm 0.4$	$1.5 \pm 0.6 \pm 0.2$

The overall agreement between the different experiments is very satisfactory. The results of calculations within the dispersion relation approach described in Sect. 3.2, using the SAID SM99K π-photoproduction multipoles of Arndt et al. [5], are included in the figure as solid lines. The parameters entering into the calculation are $\alpha_{\rm p} + \beta_{\rm p} = 13.8$, $\alpha_{\rm p} - \beta_{\rm p} = 10.6$, $\gamma_\pi^{(\rm p)} = -37.1$ and the coupling constants $g_{\pi\rm NN} F_{\pi^0\gamma\gamma}$ as given in Sect. 3.1.1 by (3.13). The dashed lines in Fig. 4.2 were obtained using $g_{\pi\rm NN} F_{\pi^0\gamma\gamma} > 0$. Thus, the results confirm that the sign of the product of the coupling constants $g_{\pi\rm NN}$ and $F_{\pi^0\gamma\gamma}$ is negative..

With the help of the dispersion relation approach, the electromagnetic polarizabilities of the proton can be extracted from the experimental cross sections. Owing to the constraints of the invariant scattering amplitudes (see Table 3.4), the difference $\alpha_{\rm p} - \beta_{\rm p}$ and the sum $\alpha_{\rm p} + \beta_{\rm p}$ may be used as parameters or, equivalently, $\alpha_{\rm p}$ and $\beta_{\rm p}$ may be used. The procedure used here was to take $\alpha_{\rm p}$ and $\beta_{\rm p}$ as free parameters and to take the constraint given by the Baldin sum rule (3.46) as a single data point. Using standard χ^2 minimization[1] when fitting the TAPS data alone (Fig. 4.2), the result obtained without the sum rule constraint is

$$\begin{aligned}
\alpha_{\rm p} &= 11.9 \pm 0.5(\text{stat.}) \mp 1.3(\text{syst.}) \,, \\
\beta_{\rm p} &= 1.2 \pm 0.7(\text{stat.}) \pm 0.3(\text{syst.}) \,.
\end{aligned} \tag{4.1}$$

The uncertainties given are the statistical (including the "random systematic" uncertainty) and the systematic uncertainty. The Baldin sum rule obtained

---

[1] In order to extract reliable uncertainties for all parameters, the program package MINUIT included in CERNlib was used.

from this result, $\alpha_\mathrm{p} + \beta_\mathrm{p} = 13.1 \pm 0.9 \mp 1.0$, is in agreement with the value in (3.46), i.e. $\alpha_\mathrm{p} + \beta_\mathrm{p} = 13.8 \pm 0.4$, within the uncertainties. The systematic uncertainties in (4.1) were obtained by rescaling the differential cross sections by $\pm 3\%$ in accordance with the common normalization uncertainty. The low-energy data of Federspiel et al. [3], MacGibbon et al. [4] and Zieger et al. [6] were treated equally; this resulted in the polarizabilities in Table 4.2.

When independent experiments are to be fitted, the procedure above is not suitable. A different way of using systematic uncertainties in fitting procedures was proposed in [8]. There, it is assumed that the systematic uncertainty is an energy-independent normalization uncertainty which can be treated like a statistical uncertainty. Therefore, according to [8], an extended χ^2 function

$$\chi^2 = \sum \left[\left(\frac{N\sigma_\mathrm{exp} - \sigma_\mathrm{theo}}{N\Delta\sigma} \right)^2 \right] + \left(\frac{N-1}{\Delta\sigma_\mathrm{sys}} \right)^2 \tag{4.2}$$

should be minimized. Here N is a normalization parameter, used to change the normalization for each data set within its systematic uncertainties $\Delta\sigma_\mathrm{sys}$. A fit to the low-energy data of [3, 4, 6] and the new TAPS data, together with the sum rule constraint of (3.46), then leads to the following result:

$$\alpha_\mathrm{p} = 12.1 \pm 0.3(\text{stat.}) \mp 0.4(\text{syst.}) \pm 0.3(\text{mod.}) \,,$$
$$\beta_\mathrm{p} = 1.6 \pm 0.4(\text{stat.}) \pm 0.4(\text{syst.}) \pm 0.4(\text{mod.}) \,, \tag{4.3}$$

where the first uncertainty denotes the statistical uncertainty, the second the systematic uncertainty and the third the model-dependent uncertainty. The results in Table 4.2 can be summarized as in Fig. 4.3, where the contours in the α_p–β_p plane for $\chi^2_\mathrm{min} + 1$ are plotted. In addition, the Baldin sum rule and the value obtained from the experiment by Zieger et al. [6] are included. The thick solid line shows the result given in (4.3) with the statistical uncertainty only.

The model-dependent uncertainties in (4.3) have been estimated by varying the main parameters entering the calculation. These are:

- the coupling constants $g_{\pi\mathrm{NN}}F_{\pi^0\gamma\gamma}$: $\Delta(g_{\pi\mathrm{NN}}F_{\pi^0\gamma\gamma}) = \pm 3.6\%$
  $\Rightarrow \Delta\alpha_\mathrm{p}(\beta_\mathrm{p}) \approx \mp 0.13(\pm 0.13)$
- the cutoff parameter Λ_π of the π^0 form factor: $\Delta\Lambda_\pi = \pm 100$ MeV
  $\Rightarrow \Delta\alpha_\mathrm{p}(\beta_\mathrm{p}) \approx \mp 0.12(\pm 0.12)$
- the strength of the M_{1+} multipole: $\Delta M_{1+} = \pm 1\%$
  $\Rightarrow \Delta\alpha_\mathrm{p}(\beta_\mathrm{p}) \approx \pm 0.10(\mp 0.10)$
- the E2/M1 ratio of the resonance multipoles: $\Delta(\text{E2/M1}) = \pm 1\%$
  $\Rightarrow \Delta\alpha_\mathrm{p}(\beta_\mathrm{p}) \approx \mp 0.08(\pm 0.08)$
- the σ mass parameter m_σ: $\Delta m_\sigma = \pm 20$ MeV
  $\Rightarrow \Delta\alpha_\mathrm{p}(\beta_\mathrm{p}) \approx \mp 0.04(\pm 0.05)$
- the backward spin polarizability $\gamma_\pi^{(\mathrm{p})}$: $\Delta\gamma_\pi^{(\mathrm{p})} = \pm 1$
  $\Rightarrow \Delta\alpha_\mathrm{p}(\beta_\mathrm{p}) \approx \pm 0.26(\mp 0.26) \,.$

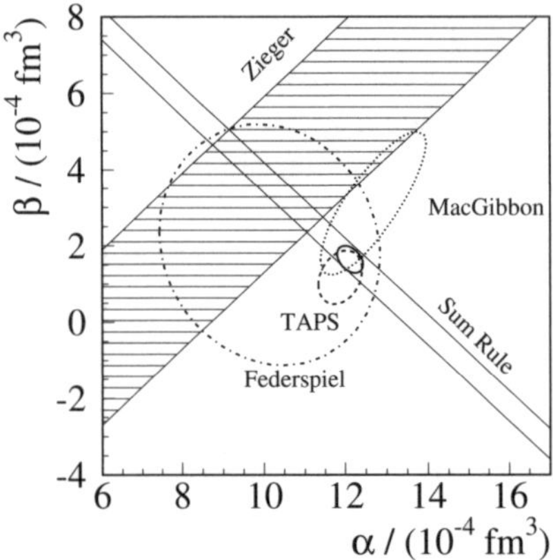

Fig. 4.3. Uncertainty contour plot in the α_p–β_p plane of the experimental results in Table 4.2 (last column), for which the uncertainties have been taken as the statistical ones only. The contours correspond to the values $\chi^2_{\min} + 1$ of the individual fits. Also shown are the sum rule constraint and the value of $\alpha_p - \beta_p$ that follows from the experiment by Zieger et al. [6]. The *thick solid line* shows the result of the global fit (4.3)

These contributions were obtained by keeping the normalization factors fixed. The total model-dependent uncertainties were then taken as the individual contributions added in quadrature. Since $\Lambda_\pi = 700$ MeV is "estimated from the axial radius of the nucleon and the size of the pion" (see [7] and references therein), the uncertainties of these quantities can be used to estimate an uncertainty of $\Delta\Lambda_\pi = \pm 100$ MeV. Another of these uncertainty estimates, i.e. $\Delta m_\sigma = \pm 20$ MeV, is based on new experimental data on Compton scattering from the proton above the Δ resonance [9, 10]. For these new data, a good description has been achieved with $m_\sigma = 600$ MeV within a range of about ± 20 MeV (see also the end of Sect. 4.2.3).

From (4.3), it follows that the difference between the electromagnetic polarizabilities is

$$\alpha_p - \beta_p = 10.5 \pm 0.9(\text{stat.} + \text{syst.}) \pm 0.7(\text{mod.}) \ . \tag{4.4}$$

This will be taken as the new global average. A comparison with the global average evaluated by MacGibbon et al. [4], $\alpha_p - \beta_p = 10.0 \pm 1.5 \pm 0.9$, exhibits the improvement achieved with the TAPS experiment. The experimental uncertainty has been reduced by a factor of almost 2/3. Agreement has also been achieved with the results of the LEGS group. The results of this group [11],

$$\alpha_p + \beta_p = 13.23 \pm 0.86(\text{stat.} + \text{syst.}) \ , \tag{4.5}$$

$$\alpha_p - \beta_p = 10.11 \pm 1.74(\text{stat.} + \text{syst.}) \ , \tag{4.6}$$

are consistent with the new results, as far as α_p and β_p are concerned.

4.1.3 Backward Spin Polarizability $\gamma_\pi^{(\mathrm{P})}$

The above results were obtained using a fixed value of the backward spin polarizability, i.e. $\gamma_\pi^{(\mathrm{P})} = -37.1$. The wide angular and energy range of the TAPS results allows one to include $\gamma_\pi^{(\mathrm{P})}$ as an additional parameter, the evaluation of which is described in the following.

The backward spin polarizability $\gamma_\pi^{(\mathrm{P})}$ is given by the non-Born contributions to the invariant amplitudes A_2 and A_5 (3.91). Only the asymptotic part of A_2 is left as a source for any additional contribution beyond π^0 exchange in the t channel. In this analysis, such a contribution has been modeled with a t-dependent term in the form of a monopole form factor,

$$A_2^{\mathrm{as}}(\nu, t) \approx A_2^{\pi^0}(t) - 2\pi m \frac{\delta\gamma_\pi^{(\mathrm{P})}}{1 - t/\Lambda^2} \,, \tag{4.7}$$

which leads to the following the substitution:

$$\gamma_\pi^{(\mathrm{P})} \rightarrow \gamma_\pi^{(\mathrm{P})} + \delta\gamma_\pi^{(\mathrm{P})} \,. \tag{4.8}$$

The parameter Λ defines the slope of the function at $t = 0$ and was chosen to be $\Lambda = 700$ MeV. By varying $\delta\gamma_\pi^{(\mathrm{P})}$, the influence of any deviation from the standard value of $\gamma_\pi^{(\mathrm{P})}$ can be investigated in terms of this ansatz.

Using the same fitting procedure as described above, the result obtained from the TAPS data alone without the sum rule constraint is

$$\begin{aligned}
\alpha_{\mathrm{p}} &= 12.2 \pm 0.8(\mathrm{stat.}) \mp 1.4(\mathrm{syst.}) \,, \\
\beta_{\mathrm{p}} &= 0.8 \pm 0.9(\mathrm{stat.}) \pm 0.5(\mathrm{syst.}) \,, \\
\gamma_\pi^{(\mathrm{P})} &= -35.9 \pm 2.3(\mathrm{stat.}) \mp 0.4(\mathrm{syst.}) \,.
\end{aligned} \tag{4.9}$$

A fit to all low-energy data including the sum rule constraint yields

$$\begin{aligned}
\alpha_{\mathrm{p}} &= 12.4 \pm 0.6(\mathrm{stat.}) \mp 0.5(\mathrm{syst.}) \pm 0.1(\mathrm{mod.}) \,, \\
\beta_{\mathrm{p}} &= 1.4 \pm 0.7(\mathrm{stat.}) \pm 0.4(\mathrm{syst.}) \pm 0.1(\mathrm{mod.}) \,, \\
\gamma_\pi^{(\mathrm{P})} &= -36.1 \pm 2.1(\mathrm{stat.}) \mp 0.4(\mathrm{syst.}) \pm 0.8(\mathrm{mod.}) \,;
\end{aligned} \tag{4.10}$$

the result for $\gamma_\pi^{(\mathrm{P})}$ can be considered as the global average obtained from low-energy Compton scattering. From these results we obtain the difference between the electromagnetic polarizabilities,

$$\alpha_{\mathrm{p}} - \beta_{\mathrm{p}} = 11.0 \pm 1.3(\mathrm{stat.} + \mathrm{syst}) \pm 0.1(\mathrm{mod.}) \,. \tag{4.11}$$

Compared with the result in (4.4), the larger statistical and systematic uncertainty is compensated by the reduced model-dependent uncertainty. The latter effect is absorbed into the large uncertainty of $\gamma_\pi^{(\mathrm{P})}$ in (4.10).

4.2 Experiments in the Δ-Resonance Region

4.2.1 Experiments with the CATS NaI(Tl) Detector

Compton scattering from the proton has been measured with the CATS NaI(Tl) detector (Sect. A.8) at laboratory angles $\theta_\gamma = 60.0°$ [12, 13], $\theta_\gamma = 130.7°$ [14] and $\theta_\gamma = 136.2°$ [15, 16]. The photon energy range covered the entire Δ-resonance region from 200 MeV to 490 MeV. The CATS NaI(Tl) detector was set up with the main collimator at a distance of about 80 cm from the target center. Thus, it covered geometrical solid angles between 20 msr and 30 msr. For all experiments, a Kapton cylinder filled with liquid hydrogen was used as the target. The target thicknesses were $(4.26 \pm 0.08) \times 10^{23}\ \mathrm{cm}^{-2}$, $(3.56 \pm 0.07) \times 10^{23}\ \mathrm{cm}^{-2}$ and $(6.42 \pm 0.13) \times 10^{23}\ \mathrm{cm}^{-2}$ for the 60°, 130° and 136° experiment, respectively. The basic features of the experiments are summarized in Table 4.3.

The 60° experiment was also equipped with an additional 2π array of BaF_2 crystals opposite to the CATS NaI(Tl) detector. This array was used to detect the low-energy photon from the asymmetric $\pi^0 \rightarrow \gamma\gamma$ decay. This allowed us to identify π^0 mesons, as well as to suppress the high-energy π^0 decay photons in the CATS NaI(Tl) detector. An overdetermination of the kinematical quantities was achieved by detecting the outgoing proton with an additional NaI(Tl) detector array. The 130° experiment made use of the CATS NaI(Tl) detector only. In case of the 136° experiment, which will be described in more detail in Sect. 5.4.1, the recoiling proton was also detected. The measured photon energy spectra were presented as a function of the missing energy $\Delta E_\gamma = \omega' - E_{\mathrm{NaI}}$. Here, ω' is the expected energy of the scattered photon (Sect. A.5), and E_{NaI} is the energy measured by the CATS NaI(Tl) detector. This kind of presentation allows one to combine several tagging channels in a single spectrum without losing the energy resolution provided by the tagger and the CATS NaI(Tl) detector. For example, two missing-energy spectra at a scattering angle of $\theta_\gamma = 130.7°$ are plotted in Fig. 4.4, showing the typical structure: (i) a peak around zero missing energy, corresponding to the elastically scattered photons, and (ii) a continuous part due to π^0 photoproduction, separated by the energy carried away by the low energy π^0 decay photon. The two regions are well separated up to incident photon energies of about 300 MeV (Fig. 4.4a). At higher energies, the two regions begin to overlap (Fig. 4.4b) so that the scattered events cannot be separated from the π^0-induced events on an event-by-event basis. Therefore, the shape of this part of the spectrum requires a simulation based on very precise angular distributions of π^0 photoproduction cross sections. After adjusting the simulated missing-energy spectra for Compton scattering and π^0 photoproduction to match the experimental spectra, the number of scattered photons can be extracted from the corresponding fit. The adjusted simulated spectra are shown in Fig. 4.4 as thick solid lines.

Table 4.3. Overview of the CATS NaI(Tl) experiments [12, 13, 14, 15, 16]

θ_γ / deg.	E_γ / MeV	$\Delta\Omega_{geom}$ / msr	ε_{tag} / %	Beam time / h
60.0 ± 5.6	200–410	30.0	50–60	150
130.7 ± 4.6	200–410	20.6	20	70
136.2 ± 5.1	200–490	20.2	55	34

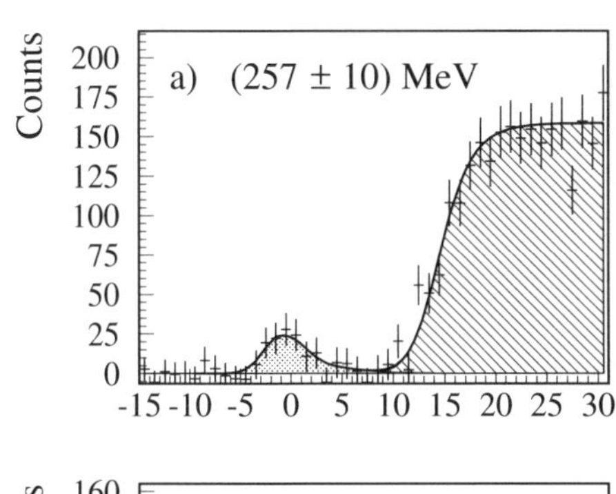

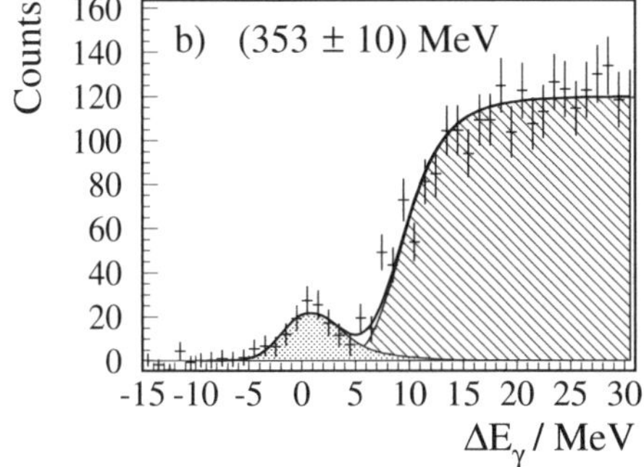

Fig. 4.4. Missing-energy spectra (*crosses*) measured with the CATS NaI(Tl) detector at $\theta_\gamma = 130.7°$ for two different incident-photon-energy bins [14]. The elastically scattered photons appear at $\Delta E_\gamma = 0$ MeV. The dotted area is the adjusted simulated response to elastically scattered photons. The adjusted simulated response to the π^0 decay photons is plotted as the *hatched area*. The sum of the two simulated responses is shown by the *thick solid line*

4.2.2 LARA Experiment

The CATS NaI(Tl) detector was able to measure at only one angle at any one setting. The **LAR**ge **A**cceptance arrangement (LARA) experiment (Sect. A.9) covered an angular range of $44°$ to $149°$ and a photon energy range extending from 200 MeV up to 800 MeV [9, 10, 17, 18]. This was achieved by detecting the scattered photons with 150 Pb glass detectors set up on a semicircle with a radius of about 2 m around the target point. In addition, the recoiling protons were detected by a wall of plastic detectors (time-of-

Table 4.4. Overview of the LARA experiment [9, 10, 17, 18]

θ_γ / deg.	E_γ / MeV	ε_{tag} / %	Beam time / h
44–149	200–800	60	150

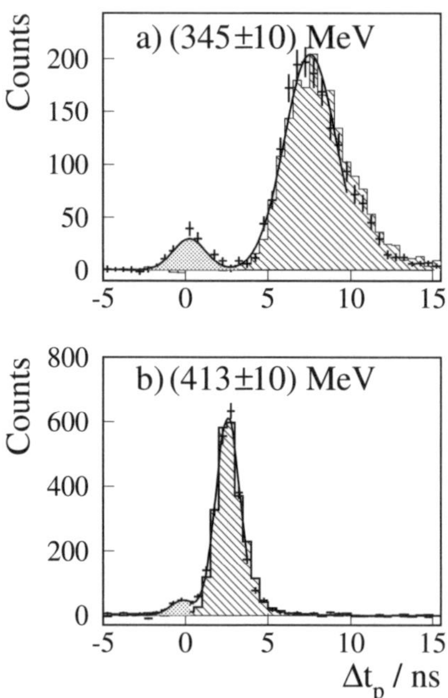

Fig. 4.5. Missing-time-of-flight spectra (*crosses*) measured with LARA at $\theta_\gamma = 65°$ for two different incident-photon-energy bins. The Compton events appear at $\Delta t_\text{p} = 0$ ns. The *solid line* is a simultaneous fit of two Gaussians to the experimental spectrum. The *dotted area* corresponds to Compton scattering. The *hatched area* is the adjusted out-of-plane spectrum, which corresponds to π^0 photoproduction events

flight (TOF) detectors) and, close to the target, with an array of multiwire proportional chambers (MWPCs) and trigger detectors (TDs). The target consisted of a Kapton cylinder filled with liquid hydrogen having a thickness of $(8.56 \pm 0.17) \times 10^{23}$ cm^{-2}. The basic features of the LARA experiment are summarized in Table 4.4.

The analysis was based on the determination of the proton recoil angle, from which the photon scattering angle θ_γ was calculated [19]. Since, in addition to the proton momentum, the photon angle could be fixed within $\pm 2.2°$ because of the geometry of the Pb glass detectors, the kinematics for Compton scattering were overdetermined. In-plane kinematics (Sect. A.5) were selected by accepting an event if a Pb glass detector and the proton trajectory lay in a plane within reasonable limits. For a fixed range of θ_γ, several neighboring TOF detectors could be hit. The missing time of flight Δt_{p} was calculated from the TOF information as the difference between the expected time of flight for Compton scattering and the measured time of flight. The resulting spectra (Fig. 4.5) exhibit the same features as in the case of the CATS NaI(Tl) experiments: (i) a peak around $\Delta t_p = 0$ from Compton scattering, and (ii) a second peak at larger time differences due to π^0 photoproduction. This second peak is part of a broad distribution and its shape at large Δt_{p} is determined by the restricted angular range of the high-energy π^0 decay photon detected. Up to 400 MeV incident photon energy, the measured spectra can be fitted with two Gaussians (solid lines in Fig. 4.5), from which the number of elastically scattered photons can be obtained.

At photon energies above 400 MeV, the two regions partly overlap (Fig. 4.5b). In this case, the π^0 events were obtained from the out-of-plane events (Sect. A.5). The ratio of in-plane to out-of-plane events was calculated with the help of a Monte Carlo simulation. As shown in Fig. 4.5, this technique works well in the Δ-energy range.

The analysis was accompanied by a detailed Monte Carlo simulation to determine the effective solid angles. In addition, a simulation was used to reproduce the measured pulse height spectra of all detectors in order to study the threshold behavior and systematic effects within the setup. Owing to the large acceptance of the detector, angular distributions provided by theoretical cross sections had to be used to generate scattered [7] and π^0-induced [5] events.

4.2.3 Results of the LARA and CATS NaI(Tl) Experiments

The differential cross sections in the Δ-resonance region measured by the CATS NaI(Tl) detector and LARA (Figs. 4.6–4.9) agree well with each other. The systematic uncertainties of the experiments have been estimated to be $\pm 4.4\% - \pm 6.6\%$ (CATS NaI(Tl), 60°), $\pm 3\%$ (CATS NaI(Tl), 130°), $\pm 4.4\%$ (CATS NaI(Tl), 136°) and $\pm 3\%$ (LARA). The interpretation in terms of dispersion relations (solid lines in Figs. 4.6–4.9) supports the correctness of the partial-wave analysis of Arndt et al. [5]. A comparison with the LEGS data [11, 23] confirms the significant difference between the MAMI and LEGS results, also seen in π photoproduction experiments [24, 25]. Therefore, only the experimental results obtained with the CATS NaI(Tl) and LARA experiments were used for further analysis.

The highly extended data base in the Δ-resonance region allows one to extract some basic properties of the proton by making use of the dispersion relation approach. These are the backward spin polarizability $\gamma_\pi^{(\mathrm{p})}$, the strength of the M1 multipole involved, and the ratio E2/M1 of the electric-quadrupole and magnetic-dipole strengths. How $\gamma_\pi^{(\mathrm{p})}$ is treated as parameter a has already been described in Sect. 4.1.2. The same formalism will be used here.

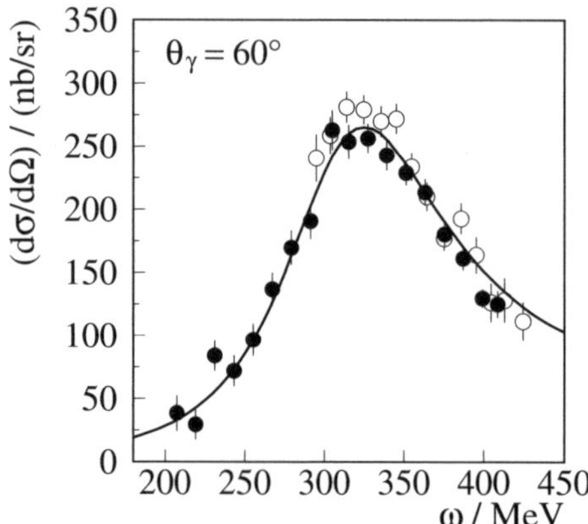

Fig. 4.6. The differential cross section in the laboratory system at $\theta_\gamma = 60.0°$ measured with the CATS NaI(Tl) detector (•) [12, 13] and measured with LARA (o) [9, 10, 19]. The *solid line* is the results of a calculation in the framework of dispersion relations as described in Sect. 3.2 using the SAID SM99K π photoproduction amplitudes of [5]

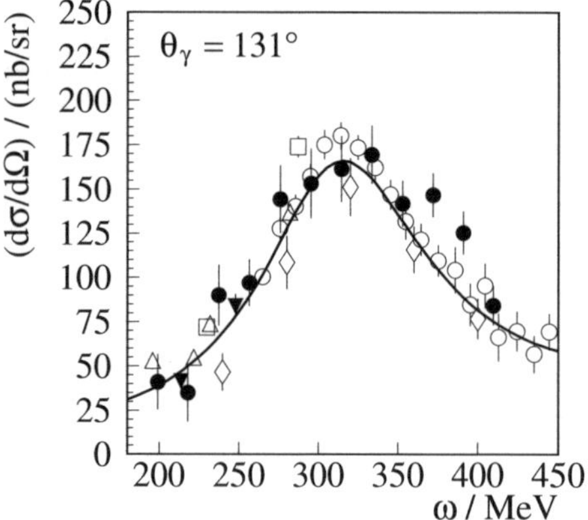

Fig. 4.7. The differential cross section in the laboratory system at $\theta_\gamma = 130.7°$ measured with the CATS NaI(Tl) detector (•) [14] and measured with LARA (o) [9, 10, 19]. The *solid line* is the results of a calculation in the framework of dispersion relations as described in Sect. 3.2 using the SAID SM99K π photoproduction amplitudes of [5]. Data are taken from [11] (□), [20] (△), [21] (▼) and [22] (◇)

The excitation of the Δ resonance occurs mainly via a magnetic dipole transition (M1). But an excitation without a change of parity is also possible via an electric quadrupole transition (E2). Whenever a quadrupole excitation is involved, one may ask whether the particle is deformed. In the case of a spin-1/2 particle, it is known that there is no static deformation observable.

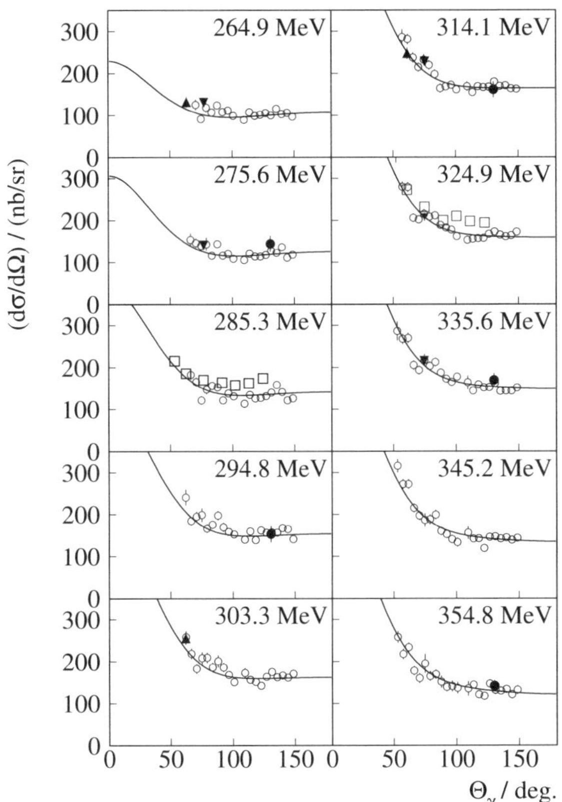

Fig. 4.8. Angular distributions of the differential cross sections in the laboratoy system measured with LARA ($\circ$). Results from other experiments are also shown: [13] ($\blacktriangle$, CATS NaI(Tl) at 60°), [14] ($\bullet$, CATS NaI(Tl) at 130°), [11, 23, 24] ($\square$, LEGS), [27] ($\blacktriangledown$, COPP at 76°). The *solid line* shows the results of a dispersion relation calculation as described in Sect. 3.2 using the SAID SM99K π photoproduction amplitudes of [5]. Continues in Fig. 4.9

But during the transition from the ground-state nucleon to the excited Δ, a deformation can be observed. Such an E2 contribution to the Δ excitation can be determined by π photoproduction. In the formalism used here, the multipoles involved have to be split into their isospin components. The E2/M1 ratio is given by the isospin-3/2 components of the imaginary parts of the M1 and E2 multipoles at the resonance energy, i.e. where the phase $\delta_{33} = 90°$. In the notation of the present formalism, these are the M_{1+} and E_{1+} multipoles. The result obtained by Beck et al. is [25, 28]

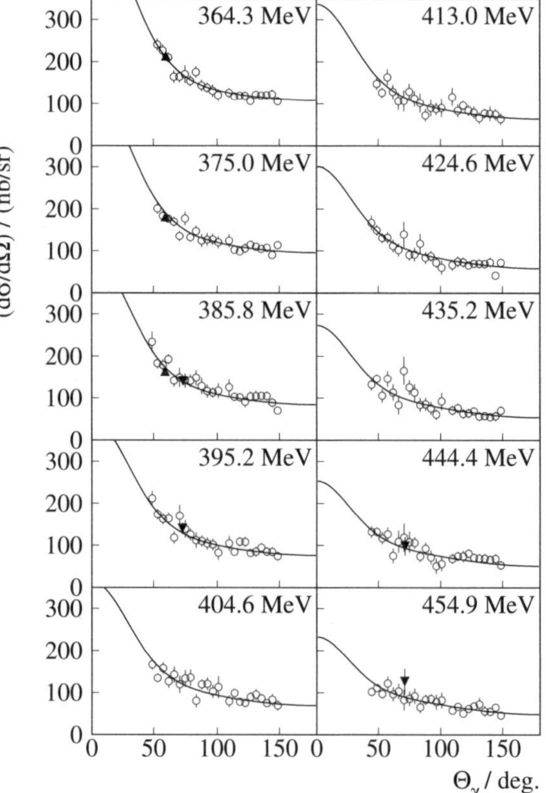

Fig. 4.9. Continuation of Fig. 4.8

$$\frac{\text{E2}}{\text{M1}} = \frac{\text{Im}\, E_{1+}^{3/2}}{\text{Im}\, M_{1+}^{3/2}} = (-2.5 \pm 0.1 \pm 0.2)\% \, . \tag{4.12}$$

The isospin decomposition of the π photoproduction multipoles is given by the isospin dependence of the two π photoproduction channels:

$$\mathcal{M}_{1+}(p\pi^0) = \mathcal{M}_{1+}^{1/2} + \frac{2}{3}\mathcal{M}_{1+}^{3/2} \, , \tag{4.13}$$

$$\mathcal{M}_{1+}(n\pi^+) = \sqrt{2}\mathcal{M}_{1+}^{1/2} - \frac{\sqrt{2}}{3}\mathcal{M}_{1+}^{3/2} \, . \tag{4.14}$$

Here $\mathcal{M} = E$, M, and it follows that

$$\mathcal{M}_{1+}^{3/2} = \mathcal{M}_{1+}(p\pi^0) - \frac{1}{\sqrt{2}}\mathcal{M}_{1+}(n\pi^+) \, . \tag{4.15}$$

From the multipole decomposition of the helicity amplitudes, as described in Sect. A.2, one may recalculate the M_{1+} and E_{1+} multipoles for both

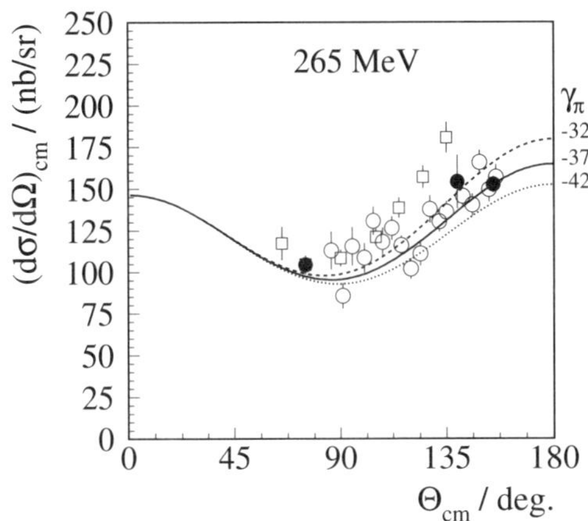

Fig. 4.10. Angular distributions of the differential cross sections in the cm system as measured with LARA ($\circ$). Results of other experiments are also shown: [13] ($\bullet$, CATS NaI(Tl) at 75°), [14] ($\bullet$, CATS NaI(Tl) at 140° and TAPS at 155°) and [11, 23, 24, 26] ($\square$, LEGS). The *solid line* shows the results of a dispersion relation calculation using the π photoproduction multipoles of Arndt et al. [5], solution SM99K, with $\gamma_\pi^{(\mathrm{P})} = -37$. The *dashed* and *dotted curves* display the results of a similar calculation with $\gamma_\pi^{(\mathrm{P})} = -32$ and $\gamma_\pi^{(p)} = -42$, respectively

reactions $\mathrm{p}(\gamma, \pi^0)\mathrm{p}$ and $\mathrm{p}(\gamma, \pi^+)\mathrm{n}$:

$$M_{1+} = \frac{1}{2}A_{1+} - \frac{3}{4}B_{1+} \ , \tag{4.16}$$

$$E_{1+} = \frac{1}{2}A_{1+} + \frac{1}{4}B_{1+} \ . \tag{4.17}$$

Inserting these into (4.15) allows one to compare the ratio with the result of Beck et al. given in (4.12)

For a more refined investigation of the M1 and E2 multipoles of Arndt et al. [5], these multipoles were parameterized into a smooth background and a resonance contribution as was done by Metcalf and Walker [29, 30]. The resonance contribution reads as

$$\mathcal{M}^{\mathrm{res}} = \mathcal{M}^{\mathrm{r}} \frac{W_0 \sqrt{\Gamma_\gamma \Gamma}}{W_0^2 - W^2 - iW_0\Gamma} \sqrt{\frac{k_0 q_0 \Gamma_0}{kp\Gamma_{\gamma0}}} \ . \tag{4.18}$$

Here, Γ and Γ_γ are the total and radiative widths, and W, k and q are the total energy, the photon momentum and π momentum, respectively, in the cm system. These latter quantities, evaluated at the resonance energy W_0, are labeled with the subscript 0. By fixing the background and rescaling the absolute value of $\mathcal{M}^{\mathrm{r}}$ by a factor f_{r}, i.e. replacing $\mathcal{M}^{\mathrm{r}}$ by $f_{\mathrm{r}}\mathcal{M}^{\mathrm{r}}$ in (4.18), new partial waves were computed. The same procedure was applied to the

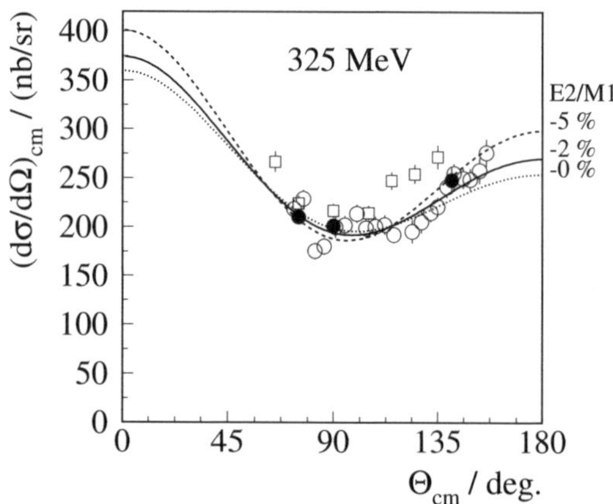

Fig. 4.11. Angular distributions of the differential cross sections in the cm system as measured with LARA ($\circ$). Results of other experiments are also shown: [13] ($\bullet$, CATS NaI(Tl) at $75°$), [14] ($\bullet$, CATS NaI(Tl) at $140°$), [27] ($\bullet$, COPP at $90°$) and [11, 23, 24, 26] ($\square$, LEGS). The *solid line* shows the results of a dispersion relation calculation using the π photoproduction multipoles of Arndt et al. [5], solution SM99K, with $E^{\mathrm{r}}_{1+}/M^{\mathrm{r}}_{1+} = -2\%$. The *dashed* and *dotted curves* display the same calculation with $E^{\mathrm{r}}_{1+}/M^{\mathrm{r}}_{1+} = -5\%$ and $E^{\mathrm{r}}_{1+}/M^{\mathrm{r}}_{1+} = 0\%$, respectively

resonance amplitude E^{r}_{1+}, except that the ratio $E^{\mathrm{r}}_{1+}/M^{\mathrm{r}}_{1+}$ was used to rescale the E_{1+} multipole. It should be mentioned that this procedure may violate the Watson theorem as far as the new multipoles are concerned. But any correction is marginal for such an investigation of the M1 strength and the E2/M1 ratio, since the background in the resonance is rather small.

The different sensitivities of the differential cross section to the parameters $\gamma^{(\mathrm{p})}_{\pi}$, f_{r} and $E^{\mathrm{r}}_{1+}/M^{\mathrm{r}}_{1+}$ allow a simultaneous fit of the measured data. Whereas f_{r} scales the entire cross section, $\gamma^{(\mathrm{p})}_{\pi}$ affects mostly the cross section at lower photon energies and at backward angles (Fig. 4.10). The ratio $E^{\mathrm{r}}_{1+}/M^{\mathrm{r}}_{1+}$ shows its influence mainly around the Δ-resonance peak at forward and backward angles (Fig. 4.11). From a minimum-χ^2 procedure applied to the LARA data alone, the following results were obtained [10]:

$$f_{\mathrm{r}} = 0.999 \pm 0.001 \,, \tag{4.19}$$

$$\frac{E^{\mathrm{r}}_{1+}}{M^{\mathrm{r}}_{1+}} = (-1.7 \pm 0.5)\% \,, \tag{4.20}$$

$$\gamma^{(\mathrm{p})}_{\pi} = -37.1 \pm 3.1 \,, \tag{4.21}$$

where the electromagnetic polarizabilities have been taken from the new global average given in (4.4). The uncertainties denote the combined statistical and systematic uncertainties. The results of calculations using this new set of parameters are shown in Figs. 4.6–4.9 by solid lines.

Including the background terms allows us to evaluate the E2/M1 ratio as defined by Beck et al. [25, 28], which depends on the isospin-3/2 components only. The result in (4.20) is then modified:

$$\frac{\text{E2}}{\text{M1}} = \frac{\text{Im}\, E_{1+}^{3/2}}{\text{Im}\, M_{1+}^{3/2}} = (-2.2 \pm 0.3(\text{stat.} + \text{syst.}) \pm 0.2(\text{mod.}))\% \,. \tag{4.22}$$

This result is in reasonable agreement with (4.12). The model-dependent uncertainty was evaluated here by varying the parameters used in the calculation within their uncertainties.

The backward spin polarizability $\gamma_\pi^{(\text{p})}$ obtained from the LARA results yields (see (4.21))

$$\gamma_\pi^{(\text{p})} = -37.1 \pm 0.6(\text{stat.} + \text{syst.}) \pm 3.0(\text{mod.}) \,. \tag{4.23}$$

The model-dependent uncertainty in (4.23) was estimated by considering the uncertainties of the parameters. Hence, the result of (4.23) confirms the global average obtained from the low-energy experiments, $\gamma_\pi^{(\text{p})} = -36.1 \pm 2.1(\text{stat.}) \mp 0.4(\text{syst.})$ (4.10).

There are now two independent experimental approaches by which $\gamma_\pi^{(\text{p})}$ has been determined, i.e. Compton scattering below and above the π production threshold. The results of both approaches are in contradiction to the result of the LEGS group, which is $\gamma_\pi^{(\text{p})} = -27.1 \pm 2.2^{+2.8}_{-2.4}$ [11]. The main reason is that the differential cross sections do not agree at large scattering angles (Fig. 4.8). This is also illustrated in Fig. 4.12 [16], where the differential cross sections in the cm system at $\theta_\gamma^* = 135°$ measured with the CATS NaI(Tl) detector and with LARA are shown. The results of the LEGS group [24, 26] clearly disagree with the CATS NaI(Tl) and LARA results. The curves in Fig. 4.12 were calculated using the most modern π photoproduction amplitudes, those of MAID2000 [31], with two different values for $\gamma_\pi^{(\text{p})}$ as given in the inset. A weighted average of all results on the backward spin polarizability of the proton obtained from Compton scattering experiments at the MAMI photon beam facility yields [16]

$$\gamma_\pi^{(\text{p})} = -38.7 \pm 1.8 \,. \tag{4.24}$$

The differential cross sections for Compton scattering from the proton measured with LARA cover an energy range up to 800 MeV [10]. Only the results in the Δ-resonance region [9] have been discussed here. The data for photon energies above 450 MeV are a crucial test of the dispersion calculation because the differential cross sections are very sensitive to the σ mass parameter in the so-called dip region, just above the Δ resonance. The σ meson was introduced to model 2π exchange in the t channel, i.e. the t-channel exchange of a σ meson. In addition, 2π photoproduction in the s channel plays an important role at these energies. Therefore, one should investigate whether or

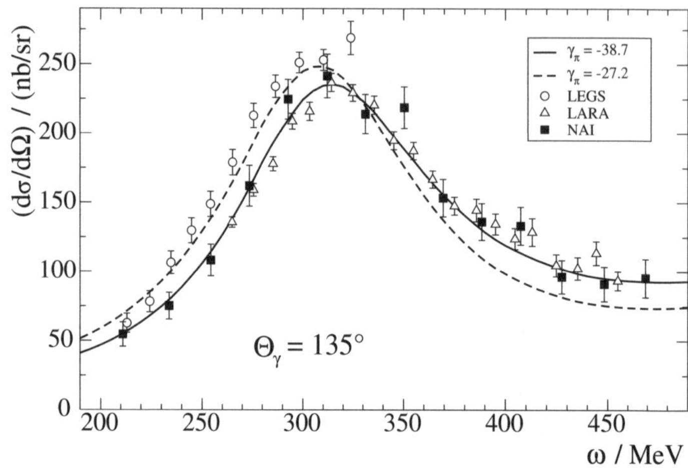

Fig. 4.12. The differential cross section in the cm system at $\theta_\gamma^* = 135°$ measured with the CATS NaI(Tl) detector (■) [15, 16], and compared with the LARA data (△) [9, 10] and LEGS data (□) [24, 26]. From [16] with permission from the American Physical Society

not this 2π channel is treated adequately in the dispersion calculation. From a detailed analysis [10], it follows that

$$m_\sigma = (589 \pm 12) \text{ MeV} . \tag{4.25}$$

The agreement with the value obtained from a phase shift analysis of $\pi\pi$ scattering, which is given in [32] as $m_\sigma^{(\pi\pi)} = (585 \pm 20)$ MeV, strongly confirms the assumption of $m_\sigma = 600$ MeV used within the calculations.

References

1. V. Olmos de León, Dissertation, Universität Mainz (2000)
2. V. Olmos de León et al., Eur. Phys. J. A **10** (2001) 207
3. F. J. Federspiel et al., Phys. Rev. Lett. **67** (1991) 1511
4. B. E. MacGibbon et al., Phys. Rev. C **52** (1995) 2097
5. R. A. Arndt et al., Phys. Rev. C **53** (1996) 430; the SAID database can be accessed via http://gwdac.phys.gwu.edu
6. A. Zieger et al., Phys. Lett. B **278** (1992) 34
7. A. I. L'vov, V. A. Petrun'kin, M. Schumacher, Phys. Rev. C **55** (1997) 359
8. G. D'Agostini, Nucl. Instrum. Methods A **346** (1994) 306
9. G. Galler et al., Phys. Lett. B **503** (2001) 245
10. S. Wolf et al., Eur. Phys. J. A **12** (2001) 231
11. J. Tonnison et al., Phys. Rev. Lett. **80** (1998) 4382
12. J. Peise et al., Phys. Lett. B **384** (1996) 37
13. A. Hünger et al., Nucl. Phys. A **620** (1997) 385

14. F. Wissmann et al., Nucl. Phys. A **660** (1999) 232; in part reprinted with permission from Elsevier
15. M. Camen, Dissertation, Universität Göttingen (2001), Cuvillier, Göttingen, 2001
16. M. Camen et al., Phys. Rev. C **65** (2002) 032202
17. G. Galler, Dissertation, Universität Göttingen (1998), Cuvillier, Göttingen, 1998
18. S. Wolf, Dissertation, Universität Göttingen (1998), Cuvillier, Göttingen, 1998
19. V. Lisin, Institute of Nuclear Research, Moscow, private communication
20. E. L. Hallin et al., Phys. Rev. C **48** (1993) 1497
21. P. S. Baranov et al., Sov. J. Nucl. Phys. **3** (1966) 791
22. H. Genzel et al., Z. Phys. A **279** (1976) 399
23. G. Blanpied et al., Phys. Rev. Lett. **76** (1996) 1023
24. G. Blanpied et al., Phys. Rev. Lett. **79** (1997) 4337
25. R. Beck et al., Phys. Rev. Lett. **78** (1997) 606
26. G. Blanpied et al., Phys. Rev. C **64** (2001) 025203
27. C. Molinari et al., Phys. Lett. B **371** (1996) 181
28. R. Beck et al., Phys. Rev. C **61** (2000) 035204
29. R. L. Walker, Phys. Rev. **182** (1969) 1729
30. W. J. Metcalf, R. L. Walker, Nucl. Phys. B **76** (1974) 253
31. D. Drechsel et al., Nucl. Phys. A **645** (1999) 145; the MAID database can be accessed via `http://www.kph.uni-mainz.de/MAID/`
32. M. Ishida, Prog. Theor. Phys. Suppl. 149 (2003) 190, *Proc. of YITP-RCNP Workshop on Chiral Restoration in Nuclear Medium*, Kyoto, Japan, 7–9 October 2002

5 Quasi-Free Compton Scattering

The previous chapters have been devoted to Compton scattering from the proton. The experimental situation concerning Compton scattering from the neutron is still very unsatisfactory. The main reason is that the free neutron is unstable and decays with a mean lifetime of about 886 s [1]. Any experiment on the neutron has to focus on nuclear targets, i.e. has to use bound neutrons. The simplest nuclei for this purpose are the deuteron, ^{2}H, and helium-3, ^{3}He. The deuteron seems to be best suited for any experiment where in addition to the produced particle the recoiling neutron is also detected, owing to the expected simplicity of a two-body problem. Without any further insight into the reactions, it is clear that a detailed theoretical calculation, which has to consider all major binding effects, has to accompany the interpretation of the experimental results.

Not only did the experiment on quasi-free Compton scattering presented below aim at the investigation of the Compton scattering process from the neutron, but it also allowed a reliable determination of the electromagnetic polarizabilities of the neutron [2], which have been the topic of an intense discussion during recent years.

5.1 Status of the Polarizabilities of the Neutron

In Fig. 5.1, the experimental status of the electric polarizability α_n of the neutron is summarized and compared with theoretical predictions. There are two different approaches to measuring α_n:[1] (i) electromagnetic scattering of low-energy neutrons in the Coulomb field of heavy nuclei , and (ii) quasi-free Compton scattering from neutrons bound in deuterons. Note here that the published results on α_n obtained from the neutron scattering experiments do not include the Schwinger term $\Delta\alpha_n = e^2\kappa_n^2/4M^3 = 0.62$ [18, 19]. After this correction is applied, the electric polarizabilities obtained from those experiments are comparable to the values obtained from Compton scattering.

Until 1991, there seemed to be a convergence of the data obtained for α_n. However, there has been a controversy about the systematic uncertainties of

---

[1] The neutron polarizabilities can also be measured by elastic Compton scattering from the deuteron [14, 15]. This method, however, is only applicable below the pion threshold [16, 17].

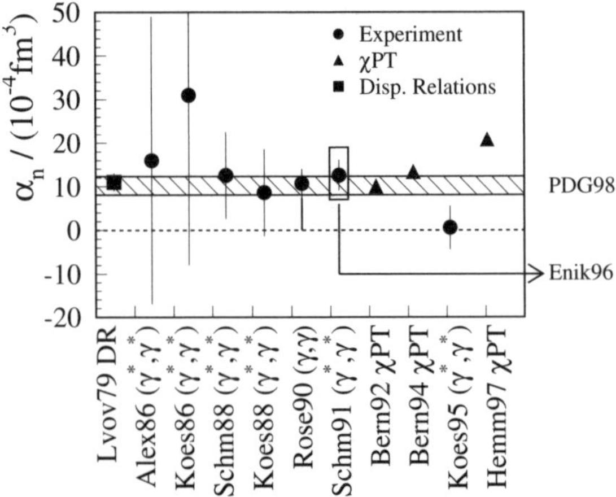

Fig. 5.1. Compilation of experimental results [3, 4, 5, 6, 7, 8, 9] and theoretical predictions [10, 11, 12, 13] for the electric polarizability of the neutron. Neutron scattering experiments are denoted by (γ^*,γ^*), and the only experiment on quasi-free Compton scattering by (γ,γ). The *hatched area* represents the value given by the Particle Data Group (PDG) [1]

neutron scattering experiments. After Schmiedmayer [3] had published his value of

$$\alpha_{\mathrm{n}} = 12.6 \pm 1.5(\text{stat.}) \pm 2.0(\text{syst.}) \,, \tag{5.1}$$

where $\Delta\alpha_{\mathrm{n}}$ has been added, Nikolenko and Popov [20, 21] raised serious doubts concerning the smallness of the quoted uncertainties. The value was obtained from the total neutron–nucleus cross section of ^{208}Pb. Taking into account resonance contributions, capture cross sections, neutron–electron and Schwinger scattering, Schmiedmayer had obtained the total scattering cross section in the energy interval 50 eV to 50 keV in the form of the expression

$$\sigma_{\mathrm{s}}(k) = 11.508(5) + 0.69(9)k - 448(3)k^2 + 9500(400)k^4 \,. \tag{5.2}$$

Here, $\sigma_{\mathrm{s}}(k)$ is given in units of barns, and $k = 2.1968 \times 10^{-4}\sqrt{E}A/(A+1)$ (where k is in fm^{-1} and E in eV, and A is the nuclear mass number) is the neutron wave number. The term linear in k corresponds to polarizability scattering. Nikolenko and Popov [20] created pseudo-experimental cross sections in accordance with the expression in (5.2), assuming different statistical uncertainties in the scattering cross section. When the first term on the right-hand side of (5.2) was reproduced with the same uncertainty, the term linear in k obtained the much larger uncertainty: 0.69(56). Nikolenko and Popov concluded that only an upper limit can be deduced from this experiment, $\alpha_{\mathrm{n}} < 20$. In 1995 Koester et al. [4] published a value, also obtained from neutron scattering experiments, which is compatible with zero:

$$\alpha_{\mathrm{n}} = 0 \pm 5 \,. \tag{5.3}$$

Again a discussion about the extraction method used to obtain the electric polarizability arose. Alexandrov [22] pointed out that the determination of α_{n} requires a statistical precision of the total neutron cross section of $\Delta\sigma/\sigma \approx 10^{-3}$. At this high level of statistical precision, it seems to be very difficult to remove possible sources of background. He also focused on the problems arising from small-angle scattering, the proper treatment of p-wave scattering and the inclusion of the term proportional to k^3 which is missing in (5.2). In 1997 Enik et al. [23] continued the discussion of the neutron scattering experiments. These authors investigated the physical interpretation of the coefficients in the expression of (5.2), concluding that there might be some problems with the cross sections measured by Schmiedmayer owing to background contributions. It was pointed out that a term proportional to k^3 has to be taken into account too. The conclusion of Enik et al.'s paper is that the systematic uncertainty in Schmiedmayer's value for α_{n} was underestimated by a factor of 3–4 and the result should be

$$\alpha_{\mathrm{n}} \sim 7\text{--}19 \,. \tag{5.4}$$

This book is not intended to contribute to the discussion of neutron scattering experiments. But by quoting the ongoing work we have illustrated that there was a need for new measurements of the neutron polarizabilities by a different method, which could be provided by quasi-free Compton scattering.

The very first experiment on quasi-free Compton scattering by neutrons bound in deuterons, carried out by Rose et al. [9], was successful in the sense that the relevant effect, i.e. coincidence events between Compton-scattered photons and recoil neutrons, was definitely identified. It was possible to extract the value $\alpha_{\mathrm{n}} = 10.7$ for the electric polarizability from the experimental data, with an upper uncertainty limit of $+3.3$. The determination of a lower uncertainty limit failed because the rather large lower uncertainty of 18% in the differential cross section did not correspond to a possible electromagnetic polarizability. In order to avoid this difficulty, the lower uncertainty limit of the differential cross section would need to have been 10% or less. At that time, the Baldin sum rule for the neutron was taken as [10]

$$\alpha_{\mathrm{n}} + \beta_{\mathrm{n}} = 15.8 \pm 0.5 \,. \tag{5.5}$$

As in the case of the proton, Babusci et al. [24] reevaluated the sum rule and obtained the value

$$\alpha_{\mathrm{n}} + \beta_{\mathrm{n}} = 14.40 \pm 0.66 \,. \tag{5.6}$$

Incorporating this new result into the evaluation of α_{n}, the following is obtained [9]:

$$\alpha_{\mathrm{n}} = 10.0^{+4.0}_{-10.0} \,. \tag{5.7}$$

A more modern analysis of the photoabsorption cross sections within the dispersion relations approach[16] leads to the value

$$\alpha_n + \beta_n = 15.2 \pm 0.5 \,. \tag{5.8}$$

From the above, it follows that new experimental attempts to determine α_n are highly desirable. One first step was made by the authors of [25], who investigated quasi-free Compton scattering from the neutron bound in the deuteron at a photon scattering angle of $135°$ integrated over an energy interval from 236 MeV to 260 MeV. Since these authors measured only one single data point, it was only possible for them to determine $\alpha_n = 12$ with a lower bound of 0. Combining their results with that one obtained by [9], they concluded that

$$\alpha_n = 7.6\text{--}14.0 \quad \text{and} \quad \beta_n = 1.2\text{--}7.6. \tag{5.9}$$

The ranges given represent the one-sigma constraints.

5.2 Theoretical Description of Quasi-Free Scattering

After the discussion concerning the unsatisfactory results of neutron scattering experiments, the only promising method for determining the electric polarizability of the neutron is quasi-free Compton scattering from the neutron bound in the deuteron,

$$\gamma d \to \gamma' np \,. \tag{5.10}$$

A detailed calculation of this reaction has been carried out by Levchuk and L'vov [26]. The main graphs contributing to this reaction are outlined in Fig. 5.2. Graphs (a) and (b) describe the quasi-free scattering from the neutron and the proton, respectively. The sum of these graphs is often referred to as the plane wave impulse approximation (PWIA). In the case of quasi-free scattering, the noninteracting nucleon behaves like a spectator. Rescattering, i.e. final-state interactions (FSI), graphs (c) and (d), and the influence of meson exchange currents (MECs) and isobar configurations (ICs) described by the graphs (e) and (f), also have to be taken into account.

Levchuk's computer code has been used for a detailed theoretical investigation of the reaction in (5.10). If not stated otherwise, all the results presented below were obtained with the deuteron wave function and np scattering amplitude for the nonrelativistic version, OBEPR, of the Bonn potential [27]. The nucleon Compton scattering amplitudes were obtained from the dispersion theory [28], as described in Sect. 3.2, using the π photoproduction multipoles of Arndt et al. [29], solution WI98K. The observable is the triple differential cross section $d^3\sigma/d\Omega_\gamma\, d\Omega_n dE_n$ for a photon–neutron pair in the final state, which shows a clear peak (neutron quasi-free peak, NQFP) around the expected energy for free scattering. Taking into account the Baldin sum rule prediction for the neutron (5.6), the cross section can be determined via the difference $\alpha_n - \beta_n$ at large scattering angles and thus α_n can be calculated.

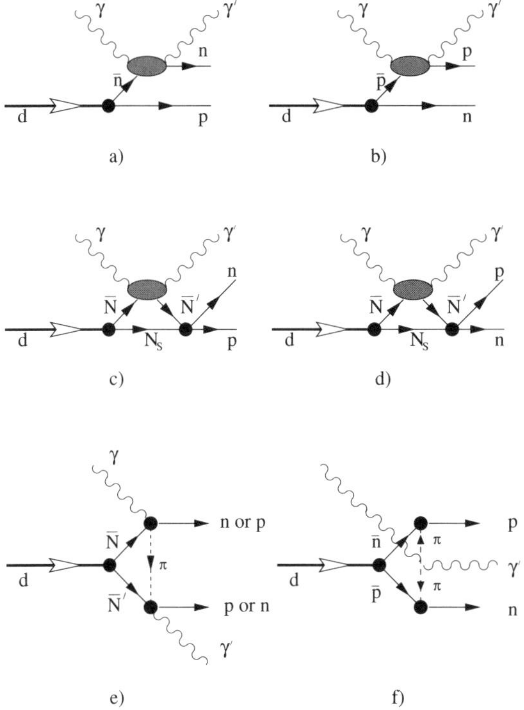

Fig. 5.2. Main graphs contributing to the reaction $\gamma d \to \gamma' np$

This cross section is plotted in Fig. 5.3 for $\omega = 100$ MeV at fixed photon scattering and neutron emission angles $\theta_\gamma = 135°$, $\vartheta_n = -20°$ (at a given photon scattering angle θ_γ, the neutron angle ϑ_n in the NQFP can be expressed approximately by $\vartheta_n \approx -(\pi - \theta_\gamma)/2$). The cross section is very small, below 1 nb/MeV/sr^2, which allows only experiments with untagged bremsstrahlung beams. On the other hand, the predicted cross sections in the center of the NQFP for $\alpha_n - \beta_n < 10$ (see also Fig. 5.6) require that the experimental uncertainties should be less than 4% in order to extract $\alpha_n - \beta_n$ with an uncertainty of about ± 2. Therefore, at photon energies below the pion threshold it is difficult but not impossible to obtain a definite value for $\alpha_n - \beta_n$.

At photon energies exceeding 200 MeV, these difficulties vanish (Figs. 5.4 and 5.5). The cross section at the NQFP is of measurable size when tagged photon beams are used. In addition, at energies between 200 MeV and 300 MeV, the cross section is rather sensitive to $\alpha_n - \beta_n$. For example, if $\alpha_n - \beta_n$ is varied from 6 to 14, the cross sections in this energy region change by more than 15%. This means that an accuracy of better than 4% must be achieved in the center of the NQFP in order to extract $\alpha_n - \beta_n$ with a precision of ± 2 (experimental only) from these experiments. An interesting energy range is that around 175 MeV, where the cross section has no sensitivity to

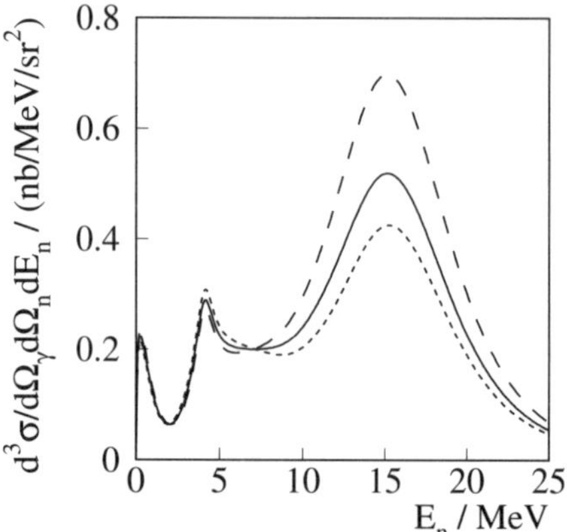

Fig. 5.3. Dependence of the triple differential cross section in the laboratory system on $\alpha_n - \beta_n$ at $\omega = 100$ MeV, $\theta_\gamma = 135°$ and $\vartheta_n = -20°$. The *dashed, solid* and *short-dashed lines* are the results of calculations with $\alpha_n - \beta_n = 20$, 10 and 0, respectively

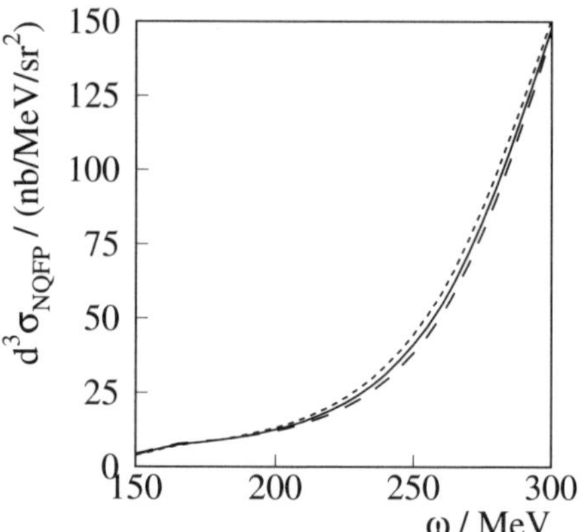

Fig. 5.4. The triple differential cross section in the center of the NQFP at $\theta_\gamma = 135°$ as a function of the incident photon energy ω, for $\alpha_n - \beta_n = 14$ (*dashed line*), 10 (*solid line*) and 6 (*short-dashed line*)

$\alpha_n - \beta_n$ at all. Near this energy. the model used is free of parameters so that measuring the differential cross section at 175 MeV would give an additional test of the reliability of the calculations of the FSI, MECs, and ICs performed in [26]. Although in Figs. 5.4 and 5.5 the results have been presented for the scattering angle $\theta_\gamma = 135°$ only, the conclusions are valid throughout the whole angular region from 120° to 180°.

Another presentation of the above results is given in Fig. 5.6. At 120 MeV, the cross section is rather flat for $\alpha_n - \beta_n < 10$. Here, the differential cross

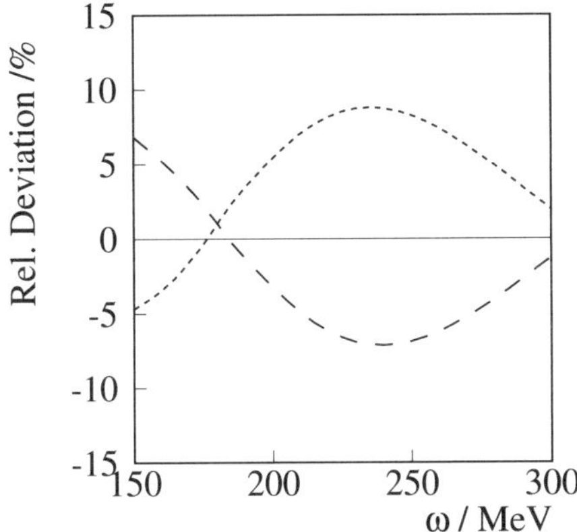

Fig. 5.5. Difference of the triple differential cross section in the center of the NQFP at $\theta_\gamma = 135°$ as a function of the incident photon energy ω, for $\alpha_n - \beta_n = 14$ (*dashed line*) and 6 (*short-dashed line*) relative to the value for $\alpha_n - \beta_n = 10$

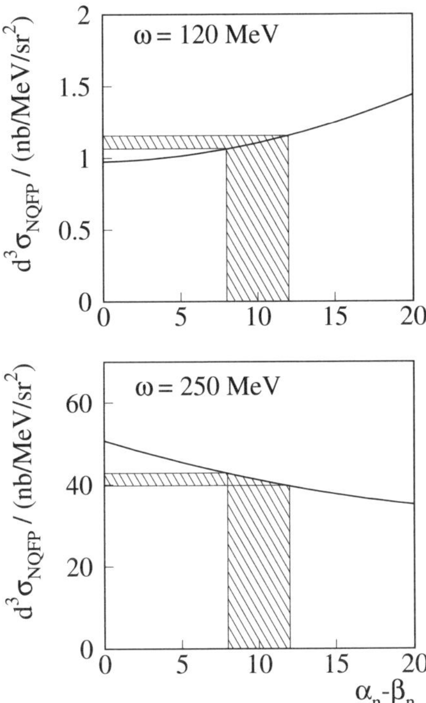

Fig. 5.6. The triple differential cross section in the center of the NQFP at $\theta_\gamma = 135°$ as a function of $\alpha_n - \beta_n$ at incident photon energies of 120 MeV (*top*) and 250 MeV (*bottom*). The *hatched areas* indicate a value of $\alpha_n - \beta_n = 10 \pm 2$ which requires the uncertainty of the measured cross section to be less than 4%

section is not very sensitive to $\alpha_n - \beta_n$. This observation also holds true at other energies below the pion threshold. Therefore, a determination of $\alpha_n - \beta_n$ is hardly possible, as experienced in the previous experiment of Rose et al. [9]. At energies above 200 MeV, this difficulty vanishes (see the bottom graph of Fig. 5.6). The cross section here is a strongly varying function of $\alpha_n - \beta_n$, which allows one to extract a definite value and upper and lower limits from the experiment.

To estimate the uncertainties arising from the theoretical modeling of the quasi-free reaction [2], the cross sections were calculated using three versions of the Bonn OBEPR potential [27, 30] and a separable approximation [31] to the Paris potential [32]. All models were found to give almost the same predictions. The relative deviation is less than 1.2% in the NQFP region at $\theta_\gamma = 135°$, reaching its maximum for the OBEPR model [27] and the Paris potential. This result can easily be understood if one takes into account that the total effect of FSI decreases from about 11% to 4% in the energy region from 200 MeV to 300 MeV [2]. The MECs and ICs contribute even less to the cross section. Therefore, any uncertainties in the parameters defining the MEC and IC contributions have a negligible impact on the extracted values of the polarizabilities.

A more serious source of uncertainties when extracting the neutron polarizabilities from the experimental data on the reaction (5.10) may be the multipole analysis of π photoproduction on the nucleon, which is necessary to evaluate the nucleon Compton scattering amplitudes used in the calculations. This question has been investigated in detail in [2], from which it follows that the accuracy of the predicted cross sections for a fixed value of $\alpha_n - \beta_n$ cannot be smaller than 5%. This in turn means that the theoretical uncertainties when $\alpha_n - \beta_n$ is determined are about ± 2.

5.3 Quasi-Free Compton Scattering from the Proton

A fundamental test of the theoretical calculations of quasi-free Compton scattering is to compare the free and quasi-free reactions on the proton. In the case of the proton, the free cross sections have been measured over a wide energy and angular range, and the theoretical description in the form of dispersion relations is well established (Sect. 4.2). There are two possibilities: (i) to use the scattering amplitudes obtained from free Compton scattering and calculate the quasi-free cross section, or (ii) to try to extract the free cross section from the measured quasi-free cross section and compare it with the measured/calculated free cross section. It is expected that both possibilities will lead to the same results.

Table 5.1. Overview of the TAPS experiment

e$^-$ Energy / MeV	θ_γ / Deg.	E_γ / MeV	$\varepsilon_{\rm tag}$ / %	Beam time / h
330	148.5 ± 12.5	200–300	32	90
855	148.5 ± 12.5	200–300	76	30

5.3.1 TAPS Experiment

Quasi-free Compton scattering from the bound proton [33, 34] in the photon energy range 200 MeV to 300 MeV has been measured with the TAPS detector system (Sect. A.7) set up at the MAMI photon beam (Sect. A.4). In this experiment, two energy settings of the electron beam, 330 MeV and 855 MeV, were analyzed. The target consisted of a Kapton cylinder 10 cm long filled with liquid deuterium. The target thickness was $N_{\rm T} = (5.41 \pm 0.11) \times 10^{23}$ cm^{-2}. Data were collected for 90 h and 30 h of beam time in the 330 MeV and 855 MeV runs, at tagging efficiencies of about 32% and 76%, respectively. The tagging efficiencies were measured with a BGO detector in the direct beam at low beam intensities. The basic features of this TAPS experiment are summarized in Table 5.1.

The scattered photons were detected within an angular range of $\pm 12.5°$ by the two blocks furthest back, positioned at $\sim 150°$ on both sides of the beam. The recoiling protons were detected in the forward wall. The protons were clearly identified with the help of pulse shape discrimination and time-of-flight analysis. Their energies were corrected for the energy loss on their way from the target to the detectors. For each photon–proton pair, a complete set of kinematical variables (Sect. A.6) was measured: E_γ, $\boldsymbol{p}'_\gamma$, $\boldsymbol{p}_{\rm p}$. Owing to the granularity of the TAPS detector, the scattering plane defined by the incoming and scattered photons could be determined very precisely. Thus, assuming quasi-free scattering from the bound proton, the missing energies of the scattered photon, ΔE_γ, and of the recoiling proton, $\Delta E_{\rm p}$, could be calculated as the difference between the expected and the measured energies. The scatter plot of these events (Fig. 5.7a) exhibits two features: (i) the scattered photons, at missing energies around zero, and (ii) a broad distribution due to quasi-free π^0 photoproduction from the proton. The two distributions strongly overlap, which is the result of the Fermi momentum distribution of the bound proton, the energy transfer due to proton–neutron rescattering and the comparatively poor photon energy resolution of the detectors (6%, to be compared with 1.5% for the CATS NaI(Tl) detector).

A projection of the spectrum in Fig. 5.7a with the cuts indicated by the dashed lines is shown in Fig. 5.7b. For further analysis, the two reactions, i.e. Compton scattering and π^0 photoproduction, were simulated separately using the GEANT code. The events were generated using the cross sections predicted in [26, 35]. Excellent agreement between the shapes of the experimental and simulated spectra was found, allowing us to adjust the simulated

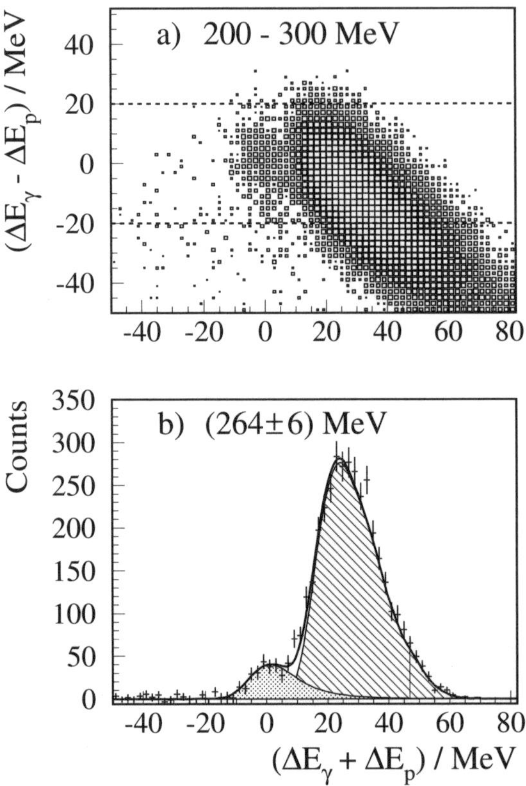

Fig. 5.7. Example of the measured events for the reaction $d(\gamma, \gamma p)n$ obtained from the run with an electron energy of 855 MeV. (a) Scatter plot of the difference $\Delta E_\gamma - \Delta E_p$ versus the sum $\Delta E_\gamma + \Delta E_p$. The latter is equivalent to the missing kinetic energy of the final particles if quasi-free scattering is assumed. All events between 200 MeV and 300 MeV incident photon energy analyzed are plotted. (b) Projection of (a) within the cut indicated by the *dashed lines* for the photon energy interval (264 ± 6) MeV. The simulated spectra from quasi-free π^0 photoproduction (*hatched area*) and quasi-free Compton scattering (*dotted area*) were simultaneously adjusted to fit the experimental spectrum. From [33] with permission from Elsevier

spectra to fit the experimental ones (Fig. 5.7b). The number of Compton-scattered photons was then obtained from the area of the corresponding fit.

The statistical precision of the experimental data for quasi-free Compton scattering from the proton does not permit one to determine the triple differential cross sections $d^3\sigma/d\Omega_{\gamma'}\, d\Omega_p\, dE_p$ as a function of the proton kinetic energy E_p. Therefore, the analysis had to start from an evaluation of the integral over this quantity with respect to the two solid angles and the proton energy. Since this integral is not independent of the geometry of the TAPS apparatus, because of geometrical constraints, it cannot be used to represent the final result of the experiment. Therefore, it was decided to carry out the analysis in such a way that the triple differential cross section $\mathrm{d}^3\sigma/\mathrm{d}\Omega_{\gamma'}\,\mathrm{d}\Omega_\mathrm{p}\,\mathrm{d}E_\mathrm{p}$ was obtained at its maximum, i.e. in the proton quasi-free peak (PQFP). This was achieved by using the GEANT code to carry out a full Monte Carlo calculation based on the predicted [26] triple differential cross section and to process the simulated data by exactly the same analyzing procedure as was applied to the experimental data.

5.3.2 Results of the TAPS Experiment

The results of the measurements of the triple differential cross section for Compton scattering from the quasi-free proton at $\theta_\gamma = 148.8°$ in the center of the PQFP are plotted in Fig. 5.8 [33]. For this experiment, the systematic uncertainties are dominated by the uncertainties of about $\pm5\%$ from the simulated detector efficiencies. Combining all systematic uncertainties in quadrature gives a total systematic uncertainty of about $\pm7\%$. Since the present measurement of quasi-free Compton scattering by the bound proton is the first such measurement, there are no data from other experiments to compare the results with. A fit to the data using the theoretical model of Levchuk and L'vov [26] with the difference between the proton polarizabilities as a parameter results in [33]

$$\alpha_\mathrm{p}-\beta_\mathrm{p} = 9.6\pm1.7\ (\text{OBEPR})\,, \quad \alpha_\mathrm{p}-\beta_\mathrm{p} = 8.5\pm1.7\ (\text{Paris potential})\,. \quad (5.11)$$

It becomes obvious that a difference of 1.1 arises between the extracted values of $\alpha_\mathrm{p} - \beta_\mathrm{p}$ depending on the use of the Bonn or the Paris potential. This difference reflects the uncertainty stemming from the N–N potential. Since no preference for either one of the N–N potentials exists, a result may be generated by averaging the values in (5.11). This leads to

$$\alpha_\mathrm{p} - \beta_\mathrm{p} = 9.1 \pm 1.7 \pm 0.6\ (\text{N–N model})\,. \quad (5.12)$$

An additional source of uncertainties arises from the multipole analysis of single-pion photoproduction from the proton, which has already been considered in Sect. 4.1.2. The result is that these model uncertainties are about ±0.9. By combining this value in quadrature with the uncertainties mentioned above, the following result has been obtained [33]:

$$\alpha_\mathrm{p} - \beta_\mathrm{p} = 9.1 \pm 1.7(\text{stat.} + \text{syst.}) \pm 1.1(\text{mod.})\,. \quad (5.13)$$

This agrees, within the uncertainties, with the new global average given in (4.4). All the results given above were obtained by combining the systematic $(\pm7\%)$ and statistical uncertainties of the cross sections in quadrature. Additional uncertainties arising from uncertainties in the incident photon energy are small enough to be neglected.

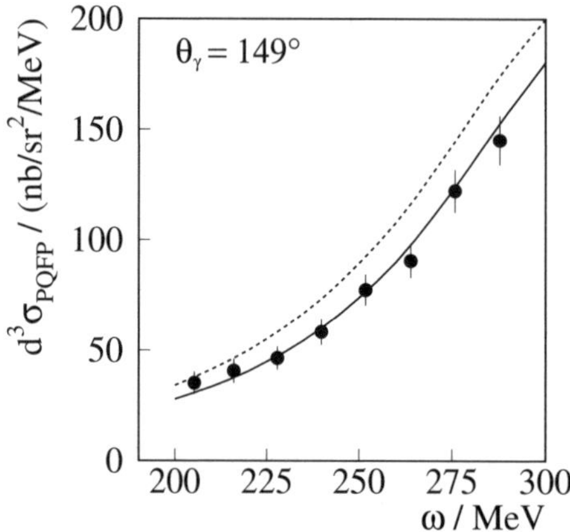

Fig. 5.8. The energy dependence of the triple differential cross section in the laboratory system for the reaction d$(\gamma, \gamma\mathrm{p})$n at $\theta_\gamma = 148.8°$ in the center of the PQFP. The uncertainties are the combined statistical and systematic uncertainties. The *solid curve* is the prediction of [26] obtained assuming the OBEPR, $\alpha_\mathrm{p} - \beta_\mathrm{p} = 9.1$, the WI98K multipoles and $\gamma_\pi^{(\mathrm{p})} = -37.6$. The *dotted curve* was obtained with $\gamma_\pi^{(\mathrm{p})} = -27.1$. The data are from the work described in this book ($\bullet$) [33, 34]. From [33] with permission from Elsevier

The experimental results are shown in comparison with the predictions of [26] in Fig. 5.8. Excellent agreement between theory and experiment for $\gamma_\pi^{(\mathrm{p})} = -37.6$ (solid line) can be seen. On the other hand, the prediction using $\gamma_\pi^{(\mathrm{p})} = -27$ clearly fails to reproduce the data (dotted line). This is also illustrated by the fact that an analogous fit to the data with $\gamma_\pi^{(\mathrm{p})} = -27$ results in $\alpha_\mathrm{p} - \beta_\mathrm{p} = 20.1\pm2.4(\mathrm{stat.+syst.})\pm1.4(\mathrm{mod.})$. This differs completely from the experimental result expressed in (4.11). In other words, these results do not give any hint of a large additional contribution to the amplitude A_2 as suggested by the LEGS group in [36]. The same conclusion has already been drawn from the LARA and TAPS results on Compton scattering from the free proton (Sect. 4.2.3).

The cross section for quasi-free Compton scattering in the center of the PQFP may be expressed in a way that displays an explicit relation between this cross section and the corresponding cross section for the free proton. This is the so-called spectator formula (see [26], Eq. (22)),

$$\frac{\mathrm{d}\sigma(\gamma\mathrm{p} \to \gamma'\mathrm{p}')}{\mathrm{d}\Omega_{\gamma'}} = \frac{(2\pi)^3}{u^2(0)} \frac{\omega\omega'}{|\boldsymbol{p}_\mathrm{p}|mE_{\gamma'}^{(\mathrm{p})2}} \frac{\mathrm{d}^3\sigma(\gamma\mathrm{d} \to \gamma'\mathrm{pn})}{\mathrm{d}\Omega_{\gamma'}\,\mathrm{d}\Omega_\mathrm{p}\,\mathrm{d}E_\mathrm{p}} , \tag{5.14}$$

where $E_{\gamma'}^{(\mathrm{p})} = p_\mathrm{p} \cdot p_{\gamma'}/m$ is the final photon energy in the rest frame of the proton, which in the case of free Compton scattering coincides with the initial photon energy in the laboratory frame, and $u(0)$ is the S-wave amplitude of the deuteron wave function at zero momentum. The difference between the

energies ω and $E_{\gamma'}^{(\mathrm{p})}$ is related to the deuteron binding energy $\Delta = 2.22$ MeV
via

$$\omega - E_{\gamma'}^{(\mathrm{p})} \simeq \Delta \left(1 + \frac{\omega}{m}\right) \tag{5.15}$$

and ranges from 2.7 MeV to 2.9 MeV in the energy region under consideration. Although the difference seems to be small, it needs to be taken into
account because of the rather strong energy dependence of the differential
cross section in this region (see [26]).

Equation (5.14) is valid for the proton pole diagram (Fig. 5.2b) contribution only. This means that when the measured cross sections are inserted
into the right-hand side of (5.14) they have to be multiplied by a factor
$f(\omega, \theta_\gamma) = \mathrm{d}^3\sigma_{\mathrm{pole}}^{\mathrm{p}} / \mathrm{d}^3\sigma_{\mathrm{tot}}$. Here, $\mathrm{d}^3\sigma_{\mathrm{pole}}^{\mathrm{p}}$ describes the contribution of the proton pole diagram (Fig. 5.2b) to the total differential cross section $\mathrm{d}^3\sigma_{\mathrm{tot}}$, in
which all of the diagrams (a)–(f) in Fig. 5.2 have been considered. The factor
was calculated in the framework of the model given in [26]. The difference
$1 - f(\omega, \theta_\gamma)$ shows the relative contribution of the background effects, which
arise mainly from FSIs and MECs. In the center of the PQFP at $\theta_\gamma^{\mathrm{lab}} = 148.8°$,
this difference ranges from –0.039 at 200 MeV to –0.023 at 290 MeV for the
OBEPR and from –0.047 to –0.030 for the Paris potential.

Using (5.14), differential cross sections for free-proton Compton scattering
can be extracted from the quasi-free data. As expected, a fit to the extracted
free data with the use of the dispersion theory [28] leads to the same values for
$\alpha_{\mathrm{p}} - \beta_{\mathrm{p}}$ as those in (5.11) and (5.12). If the most modern π photoproduction
multipoles, namely SAID SM99K or MAID2000, are used together with the
new parameters determined by the LARA experiment, the result of (5.13) is
modified. If the extracted free cross sections are fitted, and $\alpha_{\mathrm{p}} - \beta_{\mathrm{p}}$ is used
as the only parameter, the result is

$$\alpha_{\mathrm{p}} - \beta_{\mathrm{p}} = 10.0 \pm 1.6(\mathrm{stat.} + \mathrm{syst.}) \pm 1.0(\mathrm{mod.}) \, . \tag{5.16}$$

The difference between this result and (5.13) has its origin in the slightly
modified value of $\gamma_\pi^{(\mathrm{p})} = -38.7$ taken from (4.24).

Figure 5.9 presents a comparison of the cross sections for free-proton
Compton scattering extracted from the quasi-free data with the LARA results and with the predictions of dispersion theory. One can see good agreement between the predictions and experiments (solid line). The prediction
obtained using $\gamma_\pi^{(\mathrm{p})} = -27$ contradicts the data (dotted line), as expected. In
addition to the LARA experiment, only one experiment [38] exists in which
the differential cross sections for proton Compton scattering were measured
at photon scattering angles (cm system) up to 150° in the energy region
considered here, namely at 214 MeV and 249 MeV. Results of an extrapolation of the measured angular dependence of the cross sections given in [38]
to the corresponding cm angles 153.9° and 154.6° are shown in Fig. 5.9 as
open squares. The results entirely agree. Unfortunately, a direct comparison
of these results with the LEGS data in [36] is impossible, since the latter

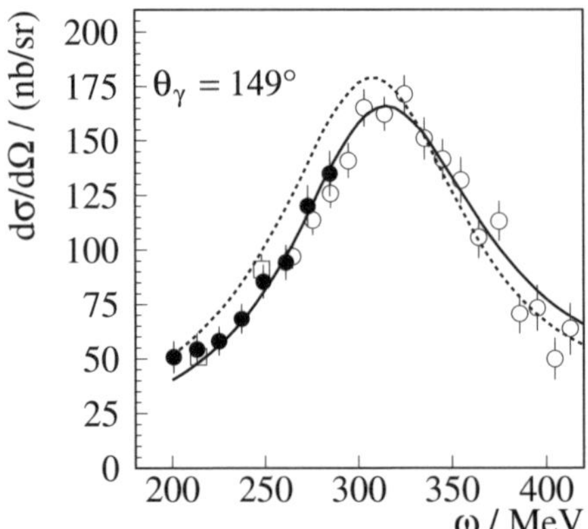

Fig. 5.9. Differential cross sections in the laboratoy system for free-proton Compton scattering at $\theta_\gamma = 148.8°$. The *filled circles* are the cross sections extracted from the quasi-free reaction (TAPS), for which the uncertainties are the combined statistical and systematic uncertainties. The *solid curve* is the prediction of the dispersion calculation [28] obtained using $\alpha_p - \beta_p = 10.0$, the SAID SM99K multipoles, and $\gamma_\pi^{(p)} = -38.7$ from (4.24). The *dotted curve* was obtained by replacing the spin polarizability by $\gamma_\pi^{(p)} = -27.0$. The additional data are from LARA ($\circ$) and [38] ($\square$)

data cover an angular range up to $\theta_\gamma^{cm} = 135°$ only. Extrapolating the angular distributions of the LEGS data to the laboratory angle of the TAPS experiment shows that the resulting cross sections exceed the TAPS results by up to about 20%, in agreement with the dashed curve in Fig. 5.9.

5.4 Quasi-Free Compton Scattering from the Neutron

After it had been shown that quasi-free Compton scattering from the proton bound in the deuteron was well understood, both theoretically and experimentally, it appeared very promising to carry out similar experiments on the neutron. The experiment of this type described below [39] aimed at a simultaneous measurement of quasi-free Compton scattering from the bound proton and neutron and, in addition, with exactly the same kinematics, Compton scattering from the free proton [40, 41, 42, 43, 44].

5.4.1 The CATS NaI(Tl)/SENECA Experiment

The experimental setup at the MAMI photon beam is outlined in Fig. 5.10. The CATS NaI(Tl) detector, with its outstanding photon energy resolution, was positioned at a scattering angle of $\theta_\gamma = 136.2°$. The recoiling neutrons/protons were detected with the SENECA detector [45] at an emission

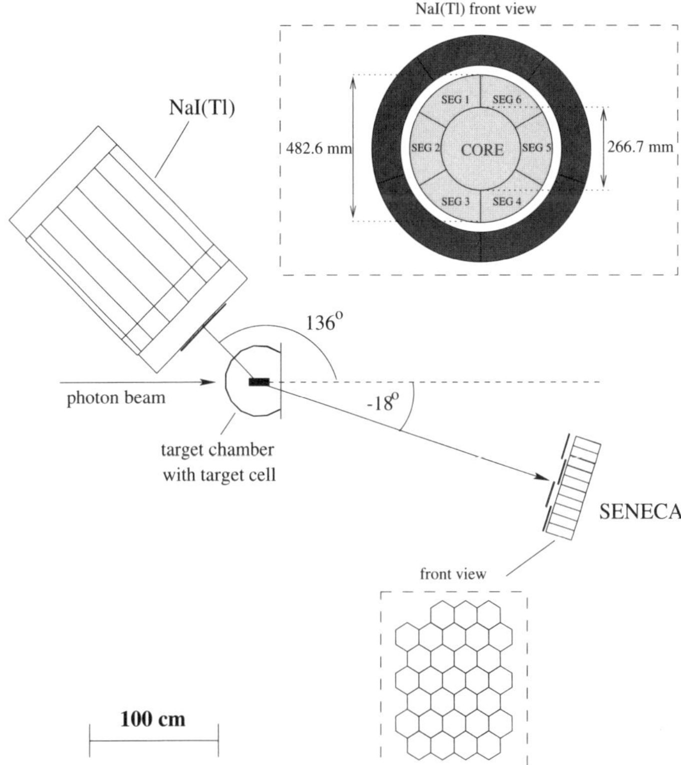

Fig. 5.10. The experimental setup to measure quasi-free Compton scattering from the bound neutron and proton. A liquid-deuterium target was used. In addition, Compton scattering from the free proton (liquid-hydrogen target) could be measured with exactly the same setup

angle of $\vartheta_N = -18°$. A distance of 250 cm was chosen to provied a compromise between a reasonable energy resolution due to the time of flight, $\Delta E_n/E_n \approx 10\%$, and the geometrical acceptance, $\Delta\Omega_n \approx 90$ msr. As the target, a 5 cm diameter × 15 cm Kapton cell filled with liquid deuterium was used.

SENECA consists of 30 hexagonally shaped cells (15 cm in diameter and 20 cm in length) mounted in a honeycomb structure. The cells are filled with liquid the scintillator NE213, which allows the separation of neutron and electromagnetically induced events via pulse-shape discrimination. But this separation only works at low neutron energies and could not be used for the purpose of the present experiment, where the energies of the recoiling neutrons were about 80 MeV. Veto detectors in front of SENECA helped to discriminate neutrons from charged particles and to identify protons. This

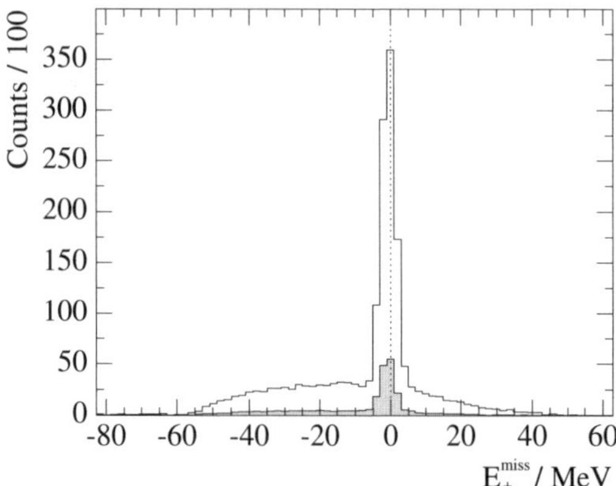

Fig. 5.11. The missing-energy spectrum of the identified π^+ meson obtained when using the CATS NaI(Tl) detector only. The *shaded area* is a similar spectrum but with SENECA in coincidence with the CATS NaI(Tl) detector

allowed us to measure quasi-free Compton scattering from the proton and neutron simultaneously [40, 42].

The time of flight of the recoiling nucleon was determined from the time difference between the scattered photon detected by the CATS NaI(Tl) detector (start signal) and the recoiling nucleon measured by SENECA (stop signal). This technique requires a good time calibration of both detector systems, which was achieved by a relative time calibration between the CATS NaI(Tl) detector and SENECA. The veto detector of the CATS NaI(Tl) detector served as the reference, since this detector was already time calibrated versus the CATS NaI(Tl) detector. Cosmic-ray-induced events were detected by coincidences between the CATS NaI(Tl) detector and the reference detector mounted underneath the CATS NaI(Tl) detector. Similar measurements were made with the reference detector mounted below SENECA. A comparison of both measurements gave a good time calibration for the time-of-flight measurements.

The tagging efficiency during the experiments was about 55%, measured by a Pb glass detector in the direct photon beam. Data were collected during 238 h of beam time. In a different run, the same target cell was filled with liquid hydrogen in order to measure Compton scattering from the free proton under exactly the same kinematical conditions [40, 41]. For this experiment, about 35 h of beam time could be used. The results of this latter experiment have been presented in Sect. 4.2.3 (see Fig. 4.12).

The neutron detection efficiency ε_n of SENECA was determined experimentally via the reaction $p(\gamma, \pi^+ n)$ with the π^+ meson detected by the CATS NaI(Tl) detector and identified as a charged particle by the veto detector. The energy measured by the CATS NaI(Tl) detector corresponds to the ki-

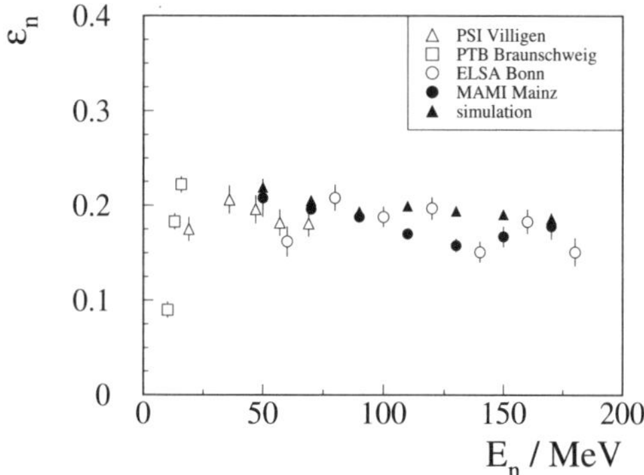

Fig. 5.12. The SENECA neutron detection efficiency ε_n as a function of the neutron kinetic energy ($\bullet$) [43, 44]. The results of efficiency measurements with monoenergetic neutron beams are also included: $\triangle$ [46], $\square$ [46, 47], $\circ$ [48]. The efficiencies evaluated from Monte Carlo simulations ($\blacktriangle$) of the CATS NaI(Tl)/SENECA experiment are also shown

netic energy of the π^+ meson after traversing various materials. Therefore, the energy loss between the reaction point and the CATS NaI(Tl) detector had to be calculated and added to the measured energy, which then gave the initial energy $E_{\pi^+}^{\mathrm{meas.}}$ of the π^+ meson. Since the initial energy of the produced meson can be evaluated from the initial photon energy and the emission angle of the π^+ meson, the calculation of the missing energy is straightforward:

$$E_{\pi^+}^{\mathrm{miss}} = E_{\pi^+}^{\mathrm{calc.}} - E_{\pi^+}^{\mathrm{meas.}} \ . \tag{5.17}$$

The measured missing-energy spectrum (Fig. 5.11) shows a clear peak around zero missing energy, corresponding to the identified π^+ mesons. A similar spectrum accumulated in coincidence with a neutral particle identified with SENECA (shaded area in Fig. 5.11), i.e. in coincidence with the recoiling neutron, is influenced only by the neutron detection efficiency of the SENECA detector. Since the same events were analyzed, the ratio of the numbers of identified π^+ mesons gives directly the neutron detection efficiency ε_n of SENECA. In Fig. 5.12 the experimentally determined neutron detection efficiency ε_n of SENECA is shown as a function of the neutron kinetic energy. The results of a detailed Monte Carlo simulation deviate significantly from the experimental results in the neutron energy range between 100 MeV and 160 MeV. This difference must be taken into account in the final simulation of the effective solid angles.

This experimental method reduced the systematic uncertainties of the neutron detection efficiency to a large extent compared with previous measurements with monoenergetic neutrons at the Paul Scherrer Institute (Villigen, Switzerland) [46], at the Physikalisch-Technische Bundesanstalt (Braun-

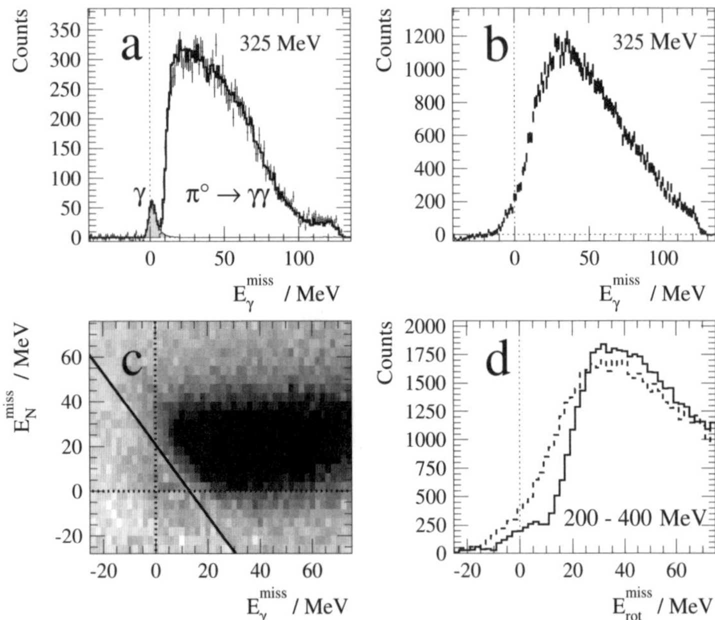

Fig. 5.13. (**a**) Missing-photon-energy spectrum for γ–proton events with a liquid-hydrogen target. The incident photon energy range is restricted to (325 ± 9) MeV. (**b**) As (**a**) but with a liquid-deuterium target. (**c**) Scatter plot of missing proton energy vs. missing photon energy when a deuteron target was used. The *thick solid line* represents the separation of Compton events around zero missing energy from (γ, π^0) events. Rotating such a plot around the zero-energy point until the thick solid line is perpendicular to the abscissa allows a one-dimensional projection. (**d**) The missing-energy spectrum before (*dashed histogram*) and after (*solid histogram*) the rotation of the scatter plot shown in (**c**)

schweig, Germany) [46, 47] and at the electron accelerator ELSA (Bonn, Germany) [48].

The quasi-free reactions on bound protons and neutrons were distinguished by the veto detectors in front of the SENECA detector. The veto detector efficiency for protons could be determined from the free-proton experiment. Here, π^0 photoproduction leads to photons detected by the CATS NaI(Tl) detector and to protons detected by SENECA. Since the protons from this reaction are emitted into a large solid angle, the detection efficiency of the entire veto system has been determined to be about 99%.

Since the momenta of the incident photon, the scattered photon and the recoiling nucleon, i.e. the proton or neutron, were measured, missing energies could be calculated assuming quasi-free kinematics (Sect. A.6). The missing energy of the scattered photon E_γ^{miss} and the missing energy of the recoiling nucleon E_N^{miss} are given by the differences between the measured and the ex-

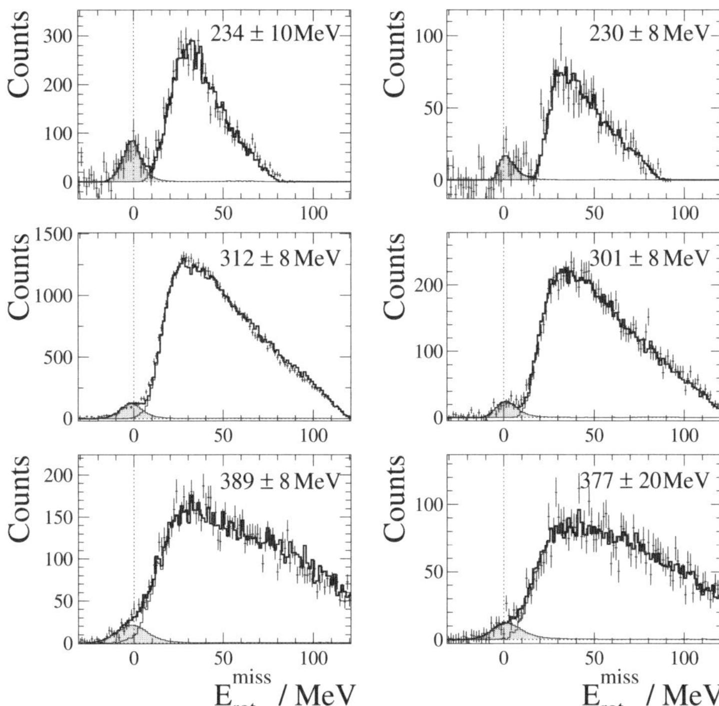

Fig. 5.14. Missing-energy spectra for different incident photon energy ranges. The *left* column shows the spectra where the recoiling nucleon was identified as a proton; the *right* column shows the spectra where it was identified as a neutron. The spectra are the projections of the two-dimensional distributions missing proton/neutron energy vs. missing photon energy (see Fig. 5.13d). The *thick solid histogram* is the result of a detailed Monte Carlo simulation of the reaction (γ, γ) around zero missing energy (*shaded area*) and (γ, π^0) at larger missing energies

pected energies. The data analysis of the quasi-free Compton scattering was optimized using the free-proton data, where a separation between Compton scattering and π^0 photoproduction events can be achieved by the CATS NaI(Tl) detector alone. The detection of the recoiling proton then improves on this separation (see Fig. 5.13a).

Here again, two distinct regions are clearly visible (Fig. 5.13): (i) the quasi-free scattered events around zero missing energy and (ii) the events from quasi-free π^0 photoproduction at larger missing energies. The strong overlap exists for the same reasons as those described in Sect. 5.3.1. A comparison between scattering from the free proton (liquid-hydrogen target) and quasi-free scattering from the proton bound in the deuteron (liquid-deuterium target) is shown in Figs. 5.13a,b. As shown in the case of the TAPS experiment described earlier (Sect. 5.3.1), a scatter plot (Fig. 5.13c) of the missing proton energy vs. the missing photon energy improves the identification of scattered events and of those originating from the (γ, π^0) reaction. This scatter plot was rotated around the center (at zero missing energies) and then projected onto a new abscissa. The resulting one-dimensional spectra clearly show the quasi-free scattered events around zero rotated missing energy $E_{\text{rot}}^{\text{miss}}$ (Fig. 5.13d).

In Fig. 5.14, the rotated missing-energy spectra where a proton was identified by SENECA (left column) and where a neutron was identified (right column) are compared. The shaded areas around zero missing energy are the simulated events of quasi-free scattered photons, adjusted to fit the experimental spectrum. The different numbers of quasi-free scattered events demonstrate the neutron detection efficiency of about 18%.

The aim of the further analysis, of course, was to determine the triple differential cross section in the center of the quasi-free peak. As in the case of the TAPS experiment (Sect. 5.3.1), this could only be achieved with the help of a detailed Monte Carlo simulation [42, 43] using very precise angular distributions of the quasi-free reactions obtained from theoretical calculations [26, 35]. In particular, a simulation of the π^0 background is essential in order to determine the number of scattered photons. The results obtained (solid histograms in Fig. 5.14) can be compared with the TAPS results (Sect. 5.3.1) as far as quasi-free scattering from the proton is concerned. The spectra in Fig. 5.14 contain all measured events. Thus, the resulting differential cross sections cover an energy range of 200 MeV to 400 MeV. Although the statistical precision of the neutron events is rather moderate, this experiment was the first one which investigated Compton scattering from the neutron covering the entire Δ-resonance region.

5.4.2 Results of the CATS NaI(Tl)/SENECA Experiment

The analysis of the measured triple differential cross sections was similar to the analysis applied to the TAPS data on quasi-free scattering from the proton (Sect. 5.3.2). The difference was that the most modern version of the amplitudes for single-π photoproduction are used, i.e. the amplitudes obtained from the MAID2000 analysis [37]. In Fig. 5.15, the cross section for quasi-free scattering from the proton is shown together with calculations performed using the diagrammatic approach by Levchuk and L'vov [26], which were mentioned in Sect. 5.2. The solid line depicts the results obtained if the MAID2000 amplitudes are used. In the center of the Δ resonance, the calculation slightly overestimates the experimental results. The long- and short-dashed curves were obtained using the solutions SM99K and SM00K from the SAID database [29] for the π photoproduction amplitudes. The difference between the results obtained using the different parameterizations enables one to estimate the theoretical uncertainty. Thus, the cross section calculated using the MAID2000 solution was rescaled by a factor of 0.93, resulting in the dash-dotted curve in Fig. 5.15. Since only two data points show

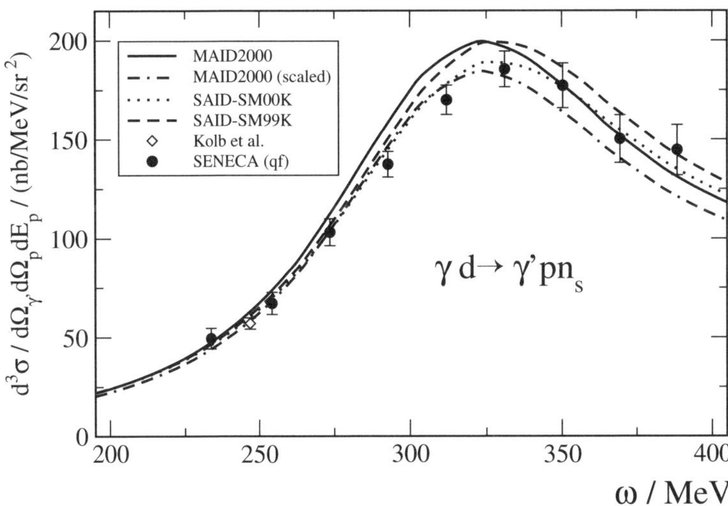

Fig. 5.15. The triple differential cross section in the center of the proton quasi-free peak at $\theta_\gamma = 136.2°$ measured by the CATS NaI(Tl)/SENECA experiment [40, 42, 43]. The result of Kolb et al. [25] at 247 MeV is also displayed

a noticable difference from the calculations, this difference may be attributed to statistical uncertainties rather than to some unknown ingredients in the theoretical description. For further analysis, the MAID2000 parameterization was used, since this solution describes the free Compton scattering from the proton best (see Fig. 4.12 in Sect. 4.2.3) [49].

The results for quasi-free Compton scattering from the neutron are shown in Fig. 5.16, together with the result of Kolb et al. [25] and the calculation by Levchuk and L'vov [26] using the MAID2000 π photoproduction multipoles. Since in the case of the free proton no deviation of the backward spin polarizability has been found, i.e. $\delta\gamma_\pi \approx 0$ (see (4.7) and (4.8)), this quantity has been kept fixed, i.e. $\gamma_\pi^{(n)} = 58.6$, as given by the dispersion calculation using the MAID2000 multipoles. Therefore, using $\alpha_n - \beta_n$ as the only parameter, the χ^2 minimization procedure leads to the following result for the neutron [43, 44]:

$$\alpha_n - \beta_n = 9.8 \pm 3.6(\text{stat.})^{+2.1}_{-1.1}(\text{syst.}) \pm 2.2(\text{mod.})\,. \qquad (5.18)$$

The systematic uncertainty has been deduced from the systematic uncertainty of the measured cross sections of $\pm 9\%$, which is given by the uncertainty of the neutron detection efficiency ($\pm 8\%$), of the target thickness ($\pm 2\%$), of the tagging efficiency ($\pm 2.5\%$) and of the analysis cuts and Monte Carlo calculations ($\pm 3\%$). The model dependent uncertainty was estimated by a comparison of the results of the fitting procedure obtained using the π photoproduction amplitudes of MAID2000 and the SAID solution SM99K.

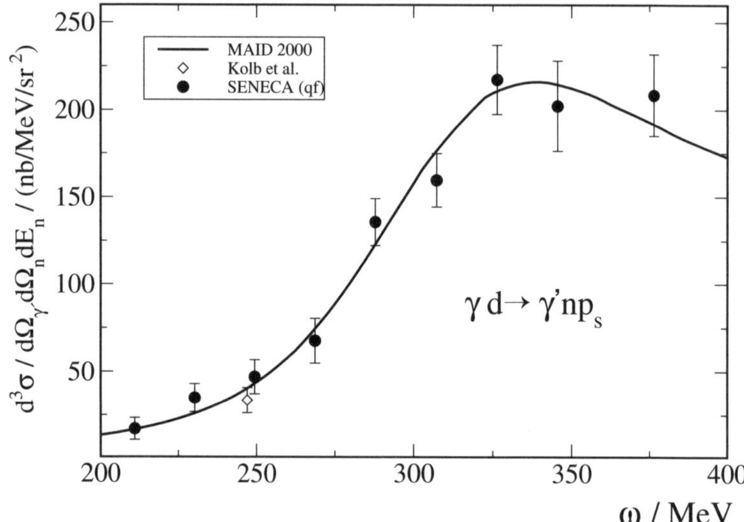

Fig. 5.16. The triple differential cross section in the center of the neutron quasi-free peak at $\theta_\gamma = 136.2°$

The result given in (5.18) can be used to extract some more information concerning the backward spin polarizability $\gamma_\pi^{(n)}$ [44]. Within the χ^2 procedure used to determine $\alpha_n - \beta_n$ the value $\gamma_\pi^{(n)} = 58.6$ was kept fixed. By changing the fixed value of $\gamma_\pi^{(n)}$, a functional dependence between the resulting value of $\alpha_n - \beta_n$ at $\chi^2_{\min}$ and the value adopted for $\gamma_\pi^{(n)}$ can be found. Since the χ^2_{min} values obtained display a broad minimum, the statistical uncertainty given in (5.18) was used to define upper and lower bounds for $\alpha_n - \beta_n$. Within the interval defined by these bounds, the valid range of $\gamma_\pi^{(n)}$ can be expressed as [44]

$$\gamma_\pi^{(n)} = 58.6 \pm 4.0 \ . \tag{5.19}$$

The Baldin sum rule for the neutron, recently reevaluated in by L'vov et al. [16], giving the value $\alpha_n + \beta_n = 15.2 \pm 0.5$, was deduced from photoabsorption data. When this value is combined with the result in (5.18), the following values for the electric and magnetic polarizabilities of the neutron are obtained:

$$\alpha_n = 12.5 \pm 1.8(\text{stat.})^{+1.1}_{-0.6}(\text{syst.}) \pm 1.1(\text{mod.}) \ ,$$
$$\beta_n = \ \ 2.7 \mp 1.8(\text{stat.})^{+0.6}_{-1.1}(\text{syst.}) \mp 1.1(\text{mod.}) \ . \tag{5.20}$$

In Sect. 5.3.2, it was explained how the differential cross sections for free scattering were extracted from the triple differential cross sections for quasi-free scattering. In the case of the proton, it was verified that below an incident photon energy of 300 MeV the extracted free cross sections agree

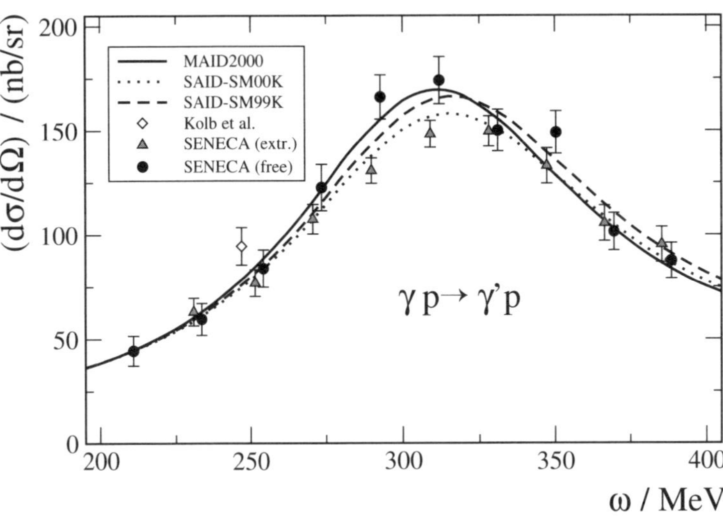

Fig. 5.17. The differential cross section for free scattering from the proton extracted from the quasi-free data (Fig. 5.15)

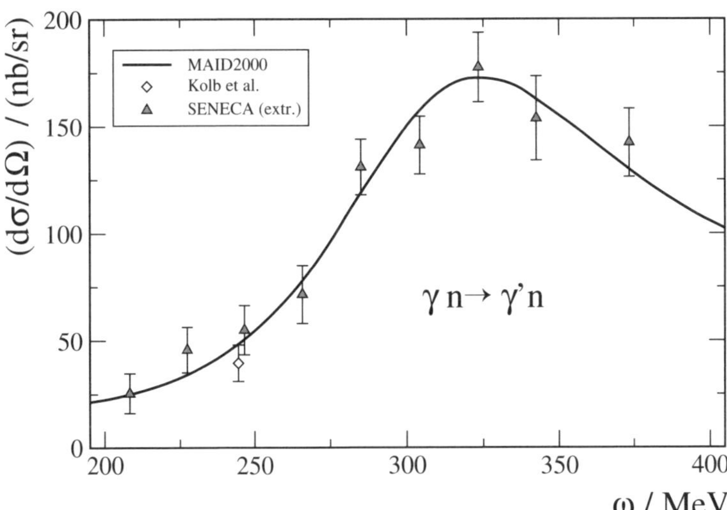

Fig. 5.18. The differential cross section for free scattering from the neutron [42, 43, 44] extracted from the quasi-free data shown in Fig. 5.16

well with the measured values. The same technique was applied to the new CATS NaI(Tl)/SENECA results. The results for the free data [40, 41] were shown in Sect. 4.2.3, Fig. 4.12. In Fig. 5.17, the extracted free cross sections of the proton are compared with the measured values. The reader should keep in mind that in both experiments, exactly the same experimental setup was used. Only the filling of the target was changed. Up to 300 MeV, there is reasonable agreement between the results. As expected from the quasi-free data, in the Δ-resonance region the extracted free cross sections are significantly lower than the measured values. The question of whether this difference is due to unknown effects in the quasi-free reaction, or is just a consequence of statistical and systematic accuracy cannot be resolved at the moment.

Looking at the systematic uncertainty of the CATS NaI(Tl)/SENECA experiments, which is $\pm 4.4\%$, it can be concluded that the description of quasi-free scattering from the proton is well under control and that the extracted differential cross sections agree fairly well with the measured values. On the basis of these findings, the extracted free cross sections for the neutron (Fig. 5.18) are the very first data on Compton scattering from the neutron which cover the entire Δ-resonance region from 200 MeV to 400 MeV. The solid line in Fig. 5.18 is the results of a calculation within the dispersion relation approach using the π photoproduction multipoles of MAID2000.

References

1. K. Hagiwara et al., Phys. Rev. D **66** (2002) 010001, *Review of Particle Physics*, Particle Data Group, http://pdg.lbl.gov
2. F. Wissmann et al., Eur. Phys. J. A **1** (1998) 193
3. J. Schmiedmayer et al., Phys. Rev. Lett. **66** (1991) 1015
4. L. Koester et al., Phys. Rev. C **51** (1995) 3363
5. Y. A. Alexandrov et al., Sov. J. Nucl. Phys. **44** (1986) 900
6. L. Koester et al., Physica B **137** (1986) 282
7. J. Schmiedmayer et al., Phys. Rev. Lett. **61** (1988) 1065
8. L. Koester et al., Z. Phys. A **329** (1988) 229
9. K. W. Rose et al., Nucl. Phys. A **514** (1990) 621
10. A. I. L'vov, V. A. Petrun'kin, S. A. Startsev, Sov. J. Nucl. Phys. **29** (1979) 651
11. V. Bernard, N. Kaiser, U.-G. Meissner, Nucl. Phys. B **373** (1992) 346
12. V. Bernard, N. Kaiser, U.-G. Meissner, Int. J. Mod. Phys. E **4** (1995) 193
13. T. R. Hemmert, B. R. Holstein, J. Kambor, Phys. Rev. D **55** (1997) 5598
14. M. I. Levchuk, A. I. L'vov, Few-Body Syst. Suppl. **9** (1995) 439
15. T. Wilbois, P. Wilhelm, H. Arenhövel, Few-Body Syst. Suppl. **9** (1995) 263
16. M. I. Levchuk, A. I. L'vov, Nucl. Phys. A **674** (2000) 449
17. D. L. Hornidge et al., Phys. Rev. Lett. **84** (2000) 2334
18. A. I. L'vov, Int. J. Mod. Phys. A **8** (1993) 5267
19. M. Bawin, S. A. Coon, Phys. Rev. C **55** (1997) 419
20. G. V. Nikolenko, A. B. Popov, Preprint E3-92-254, JINR, Dubna
21. G. V. Nikolenko, A. B. Popov, Z. Phys. A **341** (1992) 365

22. Yu. A. Alexandrov, Preprint E3-95-61, JINR, Dubna,
23. T. L. Enik et al., Phys. At. Nucl. **60** (1997) 567, Yad. Fiz. **60** (1997) 648
24. D. Babusci, G. Giordano, G. Matone, Phys. Rev. C **57** (1998) 291
25. N. R. Kolb et al., Phys. Rev. Lett. **85** (2000) 1388
26. M. I. Levchuk, A. I. L'vov, V.A. Petrun'kin, Preprint 86, FIAN, Moscow, 1986, Few-Body Syst. **16** (1994) 101
27. R. Machleidt, K. Holinde, C. Elster, Phys. Rep. **149** (1987) 1
28. A. I. L'vov, V. A. Petrun'kin, M. Schumacher, Phys. Rev. C **55** (1997) 359
29. R. A. Arndt et al., Phys. Rev. C **53** (1996) 430; the SAID database can be accessed via `http://gwdac.phys.gwu.edu`
30. R. Machleidt, Adv. Nucl. Phys. **19** (1989) 189
31. J. Haidenbauer, W. Plessas, Phys. Rev. C **30** (1984) 1822, Phys. Rev. C **32** (1985) 1424
32. M. Lacombe et al., Phys. Rev. D **12** (1975) 1495
33. F. Wissmann et al., Nucl. Phys. A **660** (1999) 232; in part reprinted with permission from Elsevier
34. V. Kuhr, Dissertation, Universität Göttingen (1998), Cuvillier, Göttingen, 1998
35. M. I. Levchuk, V. A. Petrun'kin, M. Schumacher, Z. Phys. A **355** (1996) 317
36. J. Tonnison et al., Phys. Rev. Lett. **80** (1998) 4382
37. D. Drechsel et al., Nucl. Phys. A **645** (1999) 145; the MAID database can be accessed via `http://www.kph.uni-mainz.de/MAID/`
38. P. S. Baranov et al., Sov. J. Nucl. Phys. **3** (1966) 791
39. F. Wissmann, Proposal A2-9/97, MAMI, Mainz, 1997
40. M. Camen, Dissertation, Universität Göttingen (2001), Cuvillier, Göttingen, 2001
41. M. Camen et al., Phys. Rev. C **65** (2002) 032202
42. K. Kossert, Dissertation, Universität Göttingen (2001), Cuvillier, Göttingen, 2001
43. K. Kossert et al., Phys. Rev. Lett. **88** (2002) 162301
44. K. Kossert et al., Eur. Phys. J. A **16** (2003) 259
45. G. von Edel et al., Nucl. Instrum. Methods A **365** (1993) 224
46. G. von Edel, Diploma thesis, Universität Göttingen (1992)
47. G. Galler, Diploma thesis, Universität Göttingen (1993)
48. R. Maaß, Diploma thesis, Universität Göttingen (1995)
49. S. Wolf et al., Eur. Phys. J. A **12** (2001) 231

6 Polarizabilities of the Nucleon

The experiments on Compton scattering from the proton, i.e. the TAPS experiment at low incident photon energies, and the CATS NaI(Tl) and LARA experiments in the Δ-resonance region, cover an energy range of 55 MeV to 460 MeV and photon scattering angles ranging from $44°$ to $155°$. The low-energy data measured with TAPS reveal good agreement with the experiments of [1, 2]. The global average of the electromagnetic polarizabilities of the proton has been determined with an improved precision. The difficulty arising is that the uncertainties are determined mainly by the systematic uncertainties of the experiment and by the model-dependent uncertainties. The latter are related to the π photoproduction amplitudes used within the dispersion relation approach. Thus, any further improvement would require a drastic reduction of all these uncertainties.

In the Δ-resonance region, the results of the earlier experiments [3, 4, 5] performed at the MAMI photon beam have been confirmed. Whereas at forward angles the agreement with the LEGS data [6, 7, 8] is satisfactory, there is a significant discrepancy at backward angles. The same feature is also present in experiments on π photoproduction [9, 10]. The authors of [6, 8] used their own partial-wave analysis to describe their results on π photoproduction and Compton scattering simultaneously. From that analysis, the electromagnetic polarizabilities and the backward spin polarizability were extracted as [6]

$$
\begin{aligned}
\alpha_\mathrm{p} + \beta_\mathrm{p} &= 13.23 \pm 0.86^{+0.20}_{-0.49} \,, \\
\alpha_\mathrm{p} - \beta_\mathrm{p} &= 10.11 \pm 1.74^{+1.22}_{-0.86} \,, \\
\gamma^{(\mathrm{p})}_\pi &= -27.1 \pm 2.2^{+2.8}_{-2.4} \,.
\end{aligned}
\tag{6.1}
$$

As far as α_p and β_p are concerned, the new results presented in this book and the LEGS results seem to be in agreement. But the significant difference in the backward spin polarizability, where $\gamma^{(\mathrm{p})}_\pi = -38.7 \pm 1.8$ is the combined result of the TAPS, LARA and CATS NaI(Tl) experiments [11], reflects the disagreement of the measured differential cross sections. The new experimental results presented in this book and their interpretation in terms of dispersion theory do not leave any space for an additional contribution to the amplitude A_2, as proposed by the LEGS group. This statement is also supported by a new sum rule for $\gamma_\pi^{(p)}$ evaluated by L'vov and Nathan [12], with the value $\gamma_\pi^{(p)} = -39.5 \pm 2.4$.

The experiment on quasi-free Compton scattering from the neutron bound in the deuteron presented in Sect. 5.4.1 [11, 13, 14, 15, 16] was the very first experiment on Compton scattering from the neutron covering the entire Δ-resonance region. For the first time, definite values for the electromagnetic polarizabilities of the neutron could be extracted from such an experiment. Since the earlier results obtained by Rose et al. [17] at photon energies below the π production threshold suffered from too large a statistical uncertainty, only an upper bound could be determined. The recent experiment on quasi-free scattering from the neutron by Kolb et al. [18] was the first experiment above π production threshold. The disadvantage of this experiment was that the differential cross section was measured at only one energy, i.e. at $\omega = 247$ MeV. By combining these two experimental results. Kolb et al. concluded that the valid ranges for the electromagnetic polarizabilities were

$$\alpha_n = 7.6\text{–}14.0 \ \text{ and } \ \beta_n = 1.2\text{–}7.6 \ . \tag{6.2}$$

The result $\alpha_n - \beta_n = 9.8 \pm 3.6(\text{stat.})^{+2.1}_{-1.1}(\text{syst.}) \pm 2.2(\text{mod.})$ obtained by the CATS NaI(Tl)/SENECA experiment (5.18) was a major step forward which allows a detailed discussion of the various contributions, as in the case of the proton.

For the purpose of the following discussion, the contributions to the polarizabilities of the proton and neutron were evaluated using the SAID SM99K π photoproduction multipoles. The numbers obtained are slightly modified if other multipoles are used, for example those of MAID2000. The results of the dispersion theory given in Table 6.3 and Table 6.4 were obtained by averaging the numbers obtained using the SAID-SM99K [19] and MAID2000 [20] multipoles.

6.1 Polarizabilities of the Proton

6.1.1 Contributions to the Electromagnetic Polarizabilities

The electromagnetic polarizabilities α_p and β_p were obtained from low-energy Compton scattering (Sect. 4.1.2) and the reevaluation of the Baldin sum rule (Sect. 3.1.3). The global average was deduced to be

$$\alpha_p + \beta_p = 13.8 \pm 0.4 \ , \tag{6.3}$$
$$\alpha_p - \beta_p = 10.5 \pm 0.9(\text{stat.} + \text{syst.}) \pm 0.7(\text{mod.}) \ . \tag{6.4}$$

In order to obtain more detailed insight into the internal decomposition of α_p and β_p [21], the constraints on the invariant amplitudes at forward and backward angles will be used. According to (3.74) and (3.89), one may split the contributions into an integral and an asymptotic part:

$$\alpha = \frac{1}{2}\left((\alpha+\beta)^{\mathrm{int}} + (\alpha-\beta)^{\mathrm{int}}\right)$$

$$+\frac{1}{2}\left((\alpha+\beta)^{\mathrm{as}} + (\alpha-\beta)^{\mathrm{as}}\right) , \tag{6.5}$$

$$\beta = \frac{1}{2}\left((\alpha+\beta)^{\mathrm{int}} - (\alpha-\beta)^{\mathrm{int}}\right)$$

$$+\frac{1}{2}\left((\alpha+\beta)^{\mathrm{as}} - (\alpha-\beta)^{\mathrm{as}}\right) . \tag{6.6}$$

The values calculated from the dispersion theory using the SAID SM99K π photoproduction amplitudes are

$$(\alpha_{\mathrm{p}}+\beta_{\mathrm{p}})^{\mathrm{int}} = 12.6 , \qquad (\alpha_{\mathrm{p}}+\beta_{\mathrm{p}})^{\mathrm{as}} = 1.2 ,$$

$$(\alpha_{\mathrm{p}}-\beta_{\mathrm{p}})^{\mathrm{int}} = -3.1 , \qquad (\alpha_{\mathrm{p}}-\beta_{\mathrm{p}})^{\mathrm{as}} = 13.6 . \tag{6.7}$$

This leads to the following decomposition of α_{p} and β_{p}:

$$\alpha_{\mathrm{p}} = \alpha_{\mathrm{p}}^{\mathrm{int}} + \alpha_{\mathrm{p}}^{\mathrm{as}} = 4.7 + 7.4 ,$$

$$\beta_{\mathrm{p}} = \beta_{\mathrm{p}}^{\mathrm{int}} + \beta_{\mathrm{p}}^{\mathrm{as}} = 7.8 - 6.2 . \tag{6.8}$$

The integral parts, expressed in terms of the invariant amplitudes according to Sect. 3.2, read as

$$(\alpha+\beta)^{\mathrm{int}} = -\frac{1}{2\pi}A_{3+6}^{\mathrm{int}}(0,0) = -\frac{1}{\pi^2}\int\limits_{\nu_{\mathrm{thr}}}^{1.5\,\mathrm{GeV}} \frac{\mathrm{Im}\,A_{3+6}(\nu',0)}{\nu'}\,\mathrm{d}\nu' , \tag{6.9}$$

$$(\alpha-\beta)^{\mathrm{int}} = -\frac{1}{2\pi}A_{1}^{\mathrm{int}}(0,0) = -\frac{1}{\pi^2}\int\limits_{\nu_{\mathrm{thr}}}^{1.5\,\mathrm{GeV}} \frac{\mathrm{Im}\,A_{1}(\nu',0)}{\nu'}\,\mathrm{d}\nu' . \tag{6.10}$$

Here, the variable ν coincides with the incident photon energy ω because the constraints are evaluated for $t = 0$. This leads directly to the energy dependence of the integrands of the dispersion integrals in (6.9) and (6.10), which is shown by $\mathrm{d}(\alpha,\beta)/\mathrm{d}\omega$ in Fig. 6.1 for the proton. A detailed investigation is now possible by switching on and off the individual contributions used in the evaluation of the dispersion integrals.

The main contribution to $\alpha_{\mathrm{p}}^{\mathrm{int}}$ has its origin in nonresonant single-π photoproduction: $\alpha_{\mathrm{p}}^{\mathrm{int}}(\pi) = 6.1$. This is shown in Fig. 6.1 as the dotted curves. The 2π photoproduction gives rise to $\alpha_{\mathrm{p}}^{\mathrm{int}}(2\pi) = 1.4$ (dashed curves in Fig. 6.1), which is partially canceled by the Δ-resonance contribution $\alpha_{\mathrm{p}}^{\mathrm{int}}(\Delta) = -2.8$. The solid curves in Fig. 6.1 show the sum of all contributions. One may wonder why the Δ excitation, which is purely of magnetic type, enters into the electric polarizability. This feature can be explained by rewriting (6.9) and (6.10), i.e. the imaginary parts of the amplitudes A_{3+6} and A_1, in terms

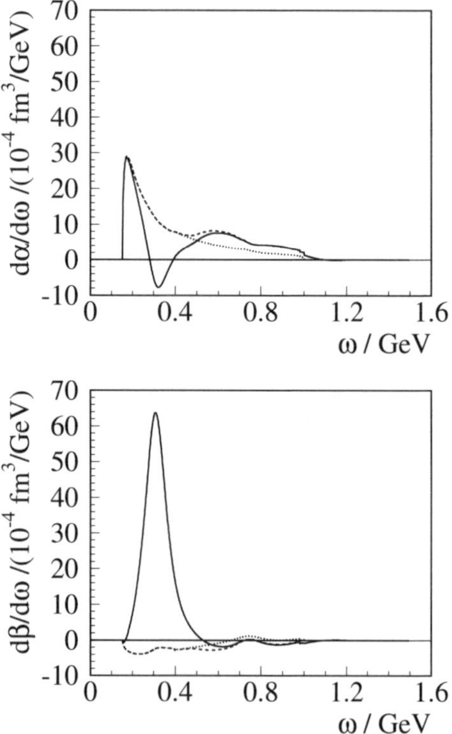

Fig. 6.1. Integrands of the dispersion integrals for $\alpha_{\mathrm{p}}^{\mathrm{int}}$ (*top*) and $\beta_{\mathrm{p}}^{\mathrm{int}}$ (*bottom*) of the proton. *Solid lines*, full calculation; *dashed lines*, no Δ contribution; *dotted lines*, no Δ contribution and no 2π contribution

of absorption cross sections. From the decomposition of the reduced helicity amplitudes τ_i into the π photoproduction multipoles, one obtains relations between the imaginary parts of A_{3+6} and A_1 and the absorption cross sections. For the sake of simplicity, only the E_{0+} (E1) and the M_{1+} (M1) multipoles are considered here. Assuming that the total absorption cross section is the sum of the electric and the magnetic absorption cross section, i.e. $\sigma_{\mathrm{tot}} = \sigma_{\mathrm{E1}} + \sigma_{\mathrm{M1}}$, the imaginary parts in (6.9) and (6.10) may be written as

$$\operatorname{Im} A_{3+6}(\nu, 0) = -\frac{1}{2\nu} \left(\sigma_{\mathrm{E1}} + \sigma_{\mathrm{M1}} \right) , \tag{6.11}$$

$$\operatorname{Im} A_1(\nu, 0) = -\frac{1}{2\nu} \left[-\frac{3}{2} \frac{s}{m^2} \sigma_{\mathrm{M1}} + \frac{\sqrt{s}}{2m} \left(2\sigma_{\mathrm{E1}} + \sigma_{\mathrm{M1}} \right) \right] . \tag{6.12}$$

If we consider only energies around the Δ resonance, i.e. $\nu/m < 1$, the expansion $s/m^2 \approx 1 + 2\nu/m$ is approximately valid. By adding and subtracting the two equations, we obtain the following decompositions:

Table 6.1. The contributions to the electromagnetic polarizabilities α_p and β_p of the proton

	π	2π	Δ	as	Sum
α_p	6.1	1.4	-2.8	7.4	12.1
β_p	-1.0	-0.6	9.4	-6.2	1.6

$$\alpha^\mathrm{int} = \frac{1}{4\pi^2} \int\limits_{\nu_\mathrm{thr}}^{1.5\,\mathrm{GeV}} \left[\left(2 + \frac{\nu}{m}\right) \sigma_\mathrm{E1} - \frac{5}{2}\frac{\nu}{m}\sigma_\mathrm{M1} \right] \frac{\mathrm{d}\nu}{\nu^2} \,, \tag{6.13}$$

$$\beta^\mathrm{int} = \frac{1}{4\pi^2} \int\limits_{\nu_\mathrm{thr}}^{1.5\,\mathrm{GeV}} \left[\left(2 + \frac{5}{2}\frac{\nu}{m}\right) \sigma_\mathrm{M1} - \frac{\nu}{m}\sigma_\mathrm{E1} \right] \frac{\mathrm{d}\nu}{\nu^2} \,. \tag{6.14}$$

The factor ν/m indicates that the magnetic-dipole excitation occurs in α^int as a recoil and retardation effect, with a large contribution due to the comparatively small proton mass. For heavy nuclei, these terms vanish and the electric polarizability is determined by the electric-dipole absorption cross section only.

As far as β_p is concerned a quite different scenario appears. Nonresonant π photoproduction and 2π photoproduction give rise to rather small effects, $\beta_\mathrm{p}^\mathrm{int}(\pi) = -1.0$ and $\beta_\mathrm{p}^\mathrm{int}(2\pi) = -0.6$. The integral part is dominated by the Δ-resonance excitation, $\beta_\mathrm{p}^\mathrm{int}(\Delta) = 9.4$. Since this excitation is related to a quark spin-flip transition, one may ascribe $\beta_\mathrm{p}^\mathrm{int}(\Delta)$ to the paramagnetic contribution. The strong diamagnetic part is then given by the pionic contributions and the large asymptotic contribution. From (6.14), it follows that the influence of the electric-dipole excitation on $\beta_\mathrm{p}^\mathrm{int}$ is much smaller than that of the magnetic-dipole excitation in $\alpha_\mathrm{p}^\mathrm{int}$. For heavy nuclei, the magnetic polarizability is completely determined by the magnetic-dipole absorption cross section.

The decomposition of α_p and β_p into the individual contributions is summarized in Table 6.1 . The conclusion drawn from the above discussion is that the electric polarizability α_p has its origin in the polarization of the pion cloud surrounding the baryonic core. This statement is based upon the large integral contribution of $\alpha_\mathrm{p}^\mathrm{int}(\pi) = 6.1$ due to nonresonant single-π photoproduction, and the asymptotic part, $\alpha_\mathrm{p}^{as} = 7.4$, which is of comparable magnitude. The latter contribution has already been related to the 2π exchange in the t channel or, in other words, to the exchange of a σ meson. In contrast to this, the magnetic polarizability β_p is strongly related to the excitation of the Δ resonance, which is expressed by the large value of $\beta_\mathrm{p}^\mathrm{int}(\Delta) = 9.4$. The large negative asymptotic contribution, $\beta_\mathrm{p}^{as} = -6.2$, can be ascribed to the diamagnetic part. Thus the net magnetic polarizability is much smaller in magnitude than the electric one.

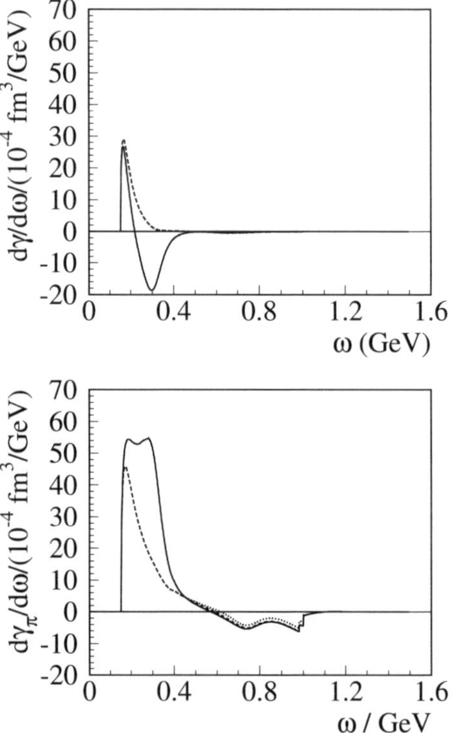

Fig. 6.2. Integrands of the dispersion integrals for $\gamma^{(\mathrm{p}),\mathrm{int}}$ (*top*) and $\gamma_\pi^{(\mathrm{p}),\mathrm{int}}$ (*bottom*) of the proton. *Solid lines*, full calculation; *dashed lines*, no Δ contribution; *dotted lines*, no Δ contribution and no 2π contribution

6.1.2 Contributions to the Spin Polarizabilities

The forward and backward spin polarizabilities, $\gamma^{(\mathrm{p})}$ and $\gamma_\pi^{(\mathrm{p})}$, respectively, can also be split into integral and asymptotic parts:

$$\gamma^{(\mathrm{p})} = \gamma^{(\mathrm{p}),\mathrm{int}} + \gamma^{(\mathrm{p}),\mathrm{as}} \;\; = -1.1 + \;\; 0.0 \,, \tag{6.15}$$

$$\gamma_\pi^{(\mathrm{p})} = \gamma_\pi^{(\mathrm{p}),\mathrm{int}} + \gamma_\pi^{(\mathrm{p}),\mathrm{as}} \;\; = \;\; 9.5 - 48.2 \,. \tag{6.16}$$

According to Sect. 3.2, the integral parts are given by

$$\gamma^{\mathrm{int}} = \;\; \frac{1}{2\pi m} A_4^{\mathrm{int}}(0,0) = \;\; \frac{1}{\pi^2 m} \int_{\nu_{\mathrm{thr}}}^{1.5\,\mathrm{GeV}} \frac{\mathrm{Im}\, A_4(\nu',0)}{\nu'} \mathrm{d}\nu' \,, \tag{6.17}$$

$$\gamma_\pi^{\mathrm{int}} = -\frac{1}{2\pi m} A_{2+5}^{\mathrm{int}}(0,0) = -\frac{1}{\pi^2 m} \int_{\nu_{\mathrm{thr}}}^{1.5\,\mathrm{GeV}} \frac{\mathrm{Im}\, A_{2+5}(\nu',0)}{\nu'} \mathrm{d}\nu' \,. \tag{6.18}$$

The integrands of the corresponding dispersion integrals are plotted in Fig. 6.2. It can be seen that the forward spin polarizability $\gamma^{(\mathrm{p})}$ is determined by nonresonant π-photoproduction, $\gamma^{(\mathrm{p}),\mathrm{int}}(\pi) = 1.8$, and a large negative amount due to the Δ-resonance, $\gamma^{(\mathrm{p}),\mathrm{int}}(\Delta) = -2.9$. The 2π photoproduction plays a minor role, $\gamma^{(\mathrm{p}),\mathrm{int}}(2\pi) \leq 0.1$, and can be neglected. The simple reason for this is the saturation of the integral at about 450 MeV, which is about 140 MeV above the 2π threshold. Experimentally, the forward spin polarizability $\gamma^{(\mathrm{p})}$ has been investigated by measuring the helicity dependence of the total photoabsorption cross section of the proton [22, 23, 24]. Since the experiments cover only a photon energy range from 200 MeV to 800 MeV, the contributions to the integral (see Sect. 3.1.3, (3.44)) from below 200 MeV and the high-energy contribution from above 800 MeV can only be estimated. The experimental result [23]

$$\gamma^{(\mathrm{p})}(200\text{ MeV–800 MeV}) = -1.87 \pm 0.08(\text{stat.}) \pm 0.10(\text{syst.}) \qquad (6.19)$$

must be corrected by $+1.04$ for the contribution from the energy region below 200 MeV. The higher-energy contribution is about -0.03 [23]. Both corrections are theoretical estimates and have to be treated rather carefully. In summary, the corrected experimental value

$$\gamma^{(\mathrm{p})} \approx -0.86 \pm 0.13(\text{stat.} + \text{syst.}), \qquad (6.20)$$

is very close to the prediction of the dispersion relation calculation based on Compton scattering from the proton given in (6.15).

From (6.16), it is obvious that the integral part of $\gamma_\pi^{(\mathrm{p})}$ is only a slight correction to the overwhelming asymptotic part $\gamma_\pi^{(\mathrm{p}),\mathrm{as}} = -48.1$. The latter has been modeled by the t-channel exchange of a neutral π meson without any major modification. The contributions of $\gamma_\pi^{(\mathrm{p}),\mathrm{int}}(\pi) = 5.1$ and $\gamma_\pi^{(\mathrm{p}),\mathrm{int}}(\Delta) = 4.8$ are almost equal and dominate the integral part. The 2π photoproduction, $\gamma_\pi^{(\mathrm{p}),\mathrm{int}}(2\pi) = -0.5$, only contributes about 5% to the integral part.

In Sect. 3.2.7, the decomposition of the four spin polarizabilities γ_{E1}, γ_{M1}, γ_{E2} and γ_{M2} into the non-Born parts of the invariant scattering amplitudes was given. An evaluation of these quantities with the help of the dispersion relation approach yields the following results for the proton:

$$\begin{aligned}
\gamma_{\mathrm{E1}}^{(\mathrm{p})} &= 7.8\ , \\
\gamma_{\mathrm{M1}}^{(\mathrm{p})} &= -9.1\ , \\
\gamma_{\mathrm{E2}}^{(\mathrm{p})} &= -9.8\ , \\
\gamma_{\mathrm{M2}}^{(\mathrm{p})} &= 12.0\ .
\end{aligned} \qquad (6.21)$$

The main contribution stems from the asymptotic part A_2^{as}, whereas A_6^{as} is negligible. A detailed analysis of the integral and the asymptotic contributions leads to

$$\gamma_{\mathrm{E1}}^{(\mathrm{p})} = -4.2 + 12.0\ (\text{int} + \text{as})\ ,$$

$$\gamma_{\mathrm{M1}}^{(\mathrm{p})} = 3.0 - 12.1 \ (\mathrm{int} + \mathrm{as}) \ ,$$
$$\gamma_{\mathrm{E2}}^{(\mathrm{p})} = 2.2 - 12.0 \ (\mathrm{int} + \mathrm{as}) \ ,$$
$$\gamma_{\mathrm{M2}}^{(\mathrm{p})} = -0.1 + 12.1 \ (\mathrm{int} + \mathrm{as}) \ . \tag{6.22}$$

To reduce the model dependence within the extraction of the spin polarizabilities, i.e. due to the asymptotic contribution due to A_2^{as}, it would be an advantage to use combinations where the asymptotic contributions vanish as in the case of the forward spin polarizability. For example, $\gamma_{\mathrm{E1}}+\gamma_{\mathrm{M1}}$, $\gamma_{\mathrm{E1}}+\gamma_{\mathrm{E2}}$, $\gamma_{\mathrm{M1}} + \gamma_{\mathrm{M2}}$ or $\gamma_{\mathrm{E2}} + \gamma_{\mathrm{M2}}$. A determination of such combinations from experiments might be achieved by double polarization experiments with polarized beam and a polarized target [25].

6.2 Polarizabilities of the Neutron

6.2.1 Contributions to the Electromagnetic Polarizabilities

The electromagnetic polarizabilities α_{n} and β_{n} of the neutron were obtained from the Baldin sum rule as evaluated by Levchuk and L'vov [26] and from quasi-free Compton scattering from the neutron bound in the deuteron (Sect. 5.4.1):

$$\alpha_{\mathrm{n}} + \beta_{\mathrm{n}} = 15.2 \pm 0.5 \ , \tag{6.23}$$
$$\alpha_{\mathrm{n}} - \beta_{\mathrm{n}} = 9.8 \pm 3.6(\mathrm{stat.})^{+2.1}_{-1.1}(\mathrm{syst.}) \pm 2.2(\mathrm{mod.}) \ . \tag{6.24}$$

As described in Sect. 6.1.1, the above values can be split into an integral and an asymptotic contribution:

$$(\alpha_{\mathrm{n}} + \beta_{\mathrm{n}})^{\mathrm{int}} = 13.9 \ , \qquad (\alpha_{\mathrm{n}} + \beta_{\mathrm{n}})^{as} = 1.3 \ ,$$

$$(\alpha_{\mathrm{n}} - \beta_{\mathrm{n}})^{int} = -1.2 \ , \qquad (\alpha_{\mathrm{n}} - \beta_{\mathrm{n}})^{as} = 11.0 \ . \tag{6.25}$$

This leads to the following decomposition of α_{n} and β_{n}:

$$\alpha_{\mathrm{n}} = \alpha_{\mathrm{n}}^{\mathrm{int}} + \alpha_{\mathrm{n}}^{\mathrm{as}} = 6.4 + 6.1 \ ,$$
$$\beta_{\mathrm{n}} = \beta_{\mathrm{n}}^{\mathrm{int}} + \beta_{\mathrm{n}}^{\mathrm{as}} = 7.6 - 4.9 \ . \tag{6.26}$$

A detailed analysis, as in the case of the proton, gives the contributions to the electromagnetic polarizabilities listed in Table 6.2. The integrands of the dispersion integrals in (6.9) and (6.10), evaluated for the neutron, are shown in Fig. 6.3. It can be seen that the main contribution to $\alpha_{\mathrm{n}}^{\mathrm{int}}$ from nonresonant single-π photoproduction is $\alpha_{\mathrm{n}}^{\mathrm{int}}(\pi) = 7.9$ (dotted curve in Fig. 6.3). The contributions from 2π photoproduction and from the Δ resonance are $\alpha_{\mathrm{n}}^{\mathrm{int}}(2\pi) = 1.2$ and $\alpha_{\mathrm{n}}^{\mathrm{int}}(\Delta) = -2.7$, respectively.

Table 6.2. The contributions to the electromagnetic polarizabilities α_n and β_n of the neutron

	π	2π	Δ	as	sum
α_n	7.9	1.2	-2.7	6.2	12.6
β_n	-0.8	-1.1	9.5	-4.9	2.7

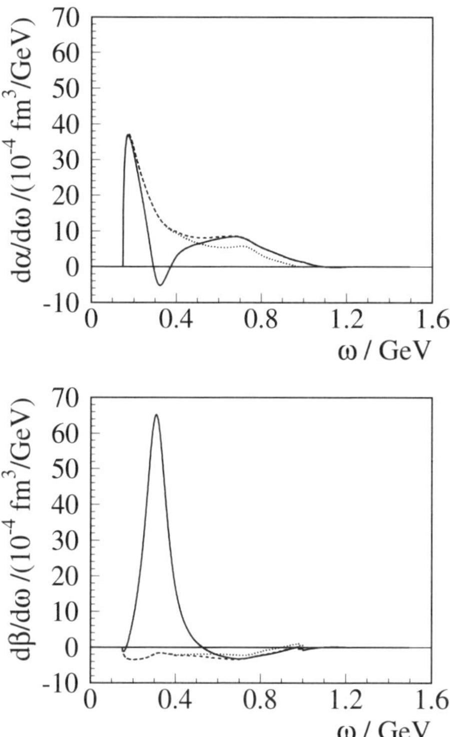

Fig. 6.3. Integrands of the dispersion integrals for α_n^{int} (*top*) and β_n^{int} (*bottom*) of the neutron. *Solid lines*, full calculation; *dashed lines*, no Δ contribution; *dotted lines*, no Δ contribution and no 2π contribution

The magnetic polarizability β_n is again dominated by the excitation of the Δ resonance (see Fig. 6.3 and Table 6.2 again), resulting in $\beta_n^{\mathrm{int}}(\Delta) = 9.5$. Nonresonant π photoproduction and 2π photoproduction give rise to $\beta_n^{\mathrm{int}}(\pi) = -0.8$ and $\beta_n^{\mathrm{int}}(2\pi) = -1.1$, respectively. Together with the asymptotic contribution $\beta_n^{\mathrm{as}} = -4.9$, the diamagnetic part greatly reduces the paramagnetic contribution.

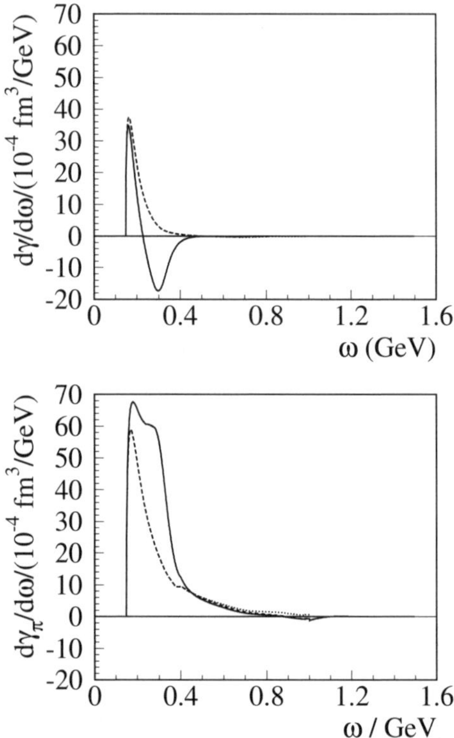

Fig. 6.4. Integrands of the dispersion integrals for $\gamma^{(n),\text{int}}$ (*top*) and $\gamma_\pi^{(n),\text{int}}$ (*bottom*) of the neutron. *Solid lines*, full calculation; *dashed lines*, no Δ-contribution; *dotted lines*, no Δ contribution and no 2π contribution

6.2.2 Contributions to the Spin Polarizabilities

The decompositions of the forward and backward spin polarizabilities $\gamma^{(n)}$ and $\gamma_\pi^{(n)}$, respectively, of the neutron are

$$\gamma^{(n)} = \gamma^{(n),\text{int}} + \gamma^{(n),\text{as}} = -0.3 + 0.0 \,, \tag{6.27}$$

$$\gamma_\pi^{(n)} = \gamma_\pi^{(n),\text{int}} + \gamma_\pi^{(n),\text{as}} = 13.4 + 45.2 \,. \tag{6.28}$$

As in the case of the proton, from the relations (6.17) and (6.18) the contributions to the dispersion integrals can be obtained. The integrands are plotted in Fig. 6.4. From an evaluation of the integrals, it is found that the integral part of the forward spin polarizability $\gamma^{(n)}$ due to nonresonant π photoproduction, i.e. $\gamma^{(n),\text{int}}(\pi) = 2.7$, is almost compensated by the Δ-resonance contribution, $\gamma^{(n),\text{int}}(\Delta) = -3.0$. The 2π contribution is now $\gamma^{(n),\text{int}}(2\pi) = 0.0$, which means, as in the proton case, that in the forward direction the 2π photoproduction amplitudes do not contribute.

The backward spin polarizability $\gamma_\pi^{(n)}$ is again dominated by the asymptotic part $\gamma_\pi^{(n),\mathrm{as}} = 45.2$, since the same exchange mechanism as for the proton is employed. The different sign as compared with the proton is due to the isospin dependence of the the Low amplitude. The contributions to the integral part $\gamma_\pi^{(n),\mathrm{int}}$ are

$$\begin{aligned}
\gamma_\pi^{(n),\mathrm{int}}(\pi) &= 8.9 \ , \\
\gamma_\pi^{(n),\mathrm{int}}(2\pi) &= -0.4 \ , \\
\gamma_\pi^{(n),\mathrm{int}}(\Delta) &= 5.0 \ .
\end{aligned} \tag{6.29}$$

An evaluation of $\gamma_{\mathrm{E1}}^{(n)}$, $\gamma_{\mathrm{M1}}^{(n)}$, $\gamma_{\mathrm{E2}}^{(n)}$ and $\gamma_{\mathrm{M2}}^{(n)}$ with the help of the dispersion relation approach yields the following results:

$$\begin{aligned}
\gamma_{\mathrm{E1}}^{(n)} &= -17.2 \ , \\
\gamma_{\mathrm{M1}}^{(n)} &= 15.0 \ , \\
\gamma_{\mathrm{E2}}^{(n)} &= 14.4 \ , \\
\gamma_{\mathrm{M2}}^{(n)} &= -12.0 \ .
\end{aligned} \tag{6.30}$$

The main contribution stems from the asymptotic part A_2^{as}, whereas A_6^{as} is negligible. A detailed analysis of the integral and asymptotic contributions leads to

$$\begin{aligned}
\gamma_{\mathrm{E1}}^{(n)} &= -5.8 - 11.4 \ (\mathrm{int} + \mathrm{as}) \ , \\
\gamma_{\mathrm{M1}}^{(n)} &= 3.8 + 11.2 \ (\mathrm{int} + \mathrm{as}) \ , \\
\gamma_{\mathrm{E2}}^{(n)} &= 3.0 + 11.4 \ (\mathrm{int} + \mathrm{as}) \ , \\
\gamma_{\mathrm{M2}}^{(n)} &= -0.8 - 11.2 \ (\mathrm{int} + \mathrm{as}) \ .
\end{aligned} \tag{6.31}$$

6.3 Theoretical Predictions of the Polarizabilities

There have been many attempts to calculate the electromagnetic and spin polarizabilities of the nucleon in various models, of which a few are given in [27, 28, 29, 30, 31, 32]. A very promising method has emerged in low-energy hadron physics: chiral perturbation theory (ChPT). The first calculation in ChPT to one-loop order was done by Bernard et al. [30], with the result

$$\alpha_\mathrm{p} = 10\beta_\mathrm{p} = 12.1 \tag{6.32}$$

A subsequent publication by the same group [31] described a calculation at $\mathcal{O}(p^4)$, where p denotes a soft momentum or the pion mass m_π; this calculation was performed in the frame of heavy-baryon ChPT (HBChPT), where the Δ contribution was also taken into account. This theory yields

$$\alpha_\mathrm{p} = 10.5 \pm 2.0 \quad \text{and} \quad \beta_\mathrm{p} = 3.5 \pm 3.6 \ , \tag{6.33}$$

Table 6.3. The polarizabilities of the proton. The column headed ChPT lists values calculated by ChPT to leading order, taken from [25, 31]. HDT is the result of a dispersion calculation by Drechsel et al. [34] using the multipoles of [35]. DR is the prediction of the dispersion relation approach described in this book. The integral and asymptotic contributions are denoted by "int" and "as". The numbers given for the integral contributions were obtained by averaging the results of dispersion calculations performed using the SAID SM99K and the MAID2000 π photoproduction multipoles. The experimental values are the numbers given with uncertainties

	ChPT			HDT			DR		
	π^0	Loop	Sum	as	int	Sum	as	int	Sum
α_p		12.1	12.1				7.4	4.7	12.1 ± 0.6
β_p		1.2	1.2				-6.2	7.8	1.6 ± 0.7
$\alpha_{\mathrm{E}\nu}$		2.2	2.2					-3.1	-3.1
$\beta_{\mathrm{M}\nu}$		3.5	3.5					8.6	8.6
$\alpha_{\mathrm{E}2}$		20.7	20.7					26.4	26.4
$\beta_{\mathrm{M}2}$		-8.9	-8.9					-20.3	-20.3
$\alpha_2 + \beta_2$		6.7	6.7					6.0	5.8 ± 0.2
$\alpha_2 - \beta_2$		-3.8	-3.8					-15.6	-15.6
$\gamma_{\mathrm{E}1}$	11.3	-5.5	5.8	11.3	-4.5	6.8	12.0	-4.2	7.8
$\gamma_{\mathrm{M}1}$	-11.3	-1.1	-12.4	-11.3	3.4	-7.9	-12.1	3.0	-9.1
$\gamma_{\mathrm{E}2}$	-11.3	1.1	-10.2	-11.3	2.3	-9.0	-12.0	2.2	-9.8
$\gamma_{\mathrm{M}2}$	11.3	1.1	12.4	11.3	-0.6	10.7	12.1	-0.1	12.0
$\gamma^{(\mathrm{p})}$		4.4	4.4		-0.6	-0.6		-1.1	-0.86 ± 0.13
$\gamma_\pi^{(\mathrm{p})}$	-45.2	4.4	-40.8	-45.2	10.8	-34.4	-48.2	9.5	-38.7 ± 1.8

where the uncertainties are the total theoretical uncertainties. As far as the spin polarizabilities are concerned, it was shown in [30] that the one-loop result for the forward spin polarizability, $\gamma^{(\mathrm{p})} = 2.2$, is in contradiction to the expected value as obtained from the partial-wave analysis of Arndt et al. [19] (see also Sect. A.2), especially in its sign.

Including the Δ resonance in some phenomenological way leads to a large negative contribution of $\gamma^{(\mathrm{p})}(\Delta) = -3.7$. The sum of both contributions gives the prediction

$$\gamma^{(\mathrm{p})} = -1.5 \,. \tag{6.34}$$

A comparison with the result obtained by the dispersion relation approach given in (6.15), $\gamma^{(\mathrm{p})} = -1.1$, indicates that including the Δ contribution leads in the right direction.

The importance of the Δ resonance led Hemmert et al. [32] to add the Δ contribution by means of a further expansion parameter, the nucleon–Δ mass difference $\epsilon = m_\Delta - m$ ("small-scale expansion"). The result to the order $\mathcal{O}(\epsilon^3)$ [32, 33] is

$$\alpha_\mathrm{p} = 16.4 \,, \qquad \beta_\mathrm{p} = \quad 9.1 \,,$$
$$\gamma^{(\mathrm{p})} = \quad 2.0 \,, \qquad \gamma_\pi^{(\mathrm{p})} = -36.7 \,. \tag{6.35}$$

Table 6.4. As in Table 6.3 but for the neutron

	π^0	ChPT Loop	Sum	HDT as	int	Sum	DR as	int	Sum
α_{n}		12.1	12.1				6.1	6.4	12.5 ± 2.4
β_{n}		1.2	1.2				-4.9	7.6	2.7 ± 2.4
$\alpha_{\mathrm{E}\nu}$		2.2	2.2					-1.4	-1.4
$\beta_{\mathrm{M}\nu}$		3.5	3.5					8.4	8.4
$\alpha_{\mathrm{E}2}$		20.7	20.7					24.1	24.1
$\beta_{\mathrm{M}2}$		-8.9	-8.9					-20.0	-20.0
$\alpha_2 + \beta_2$		6.7	6.7					7.3	7.3
$\alpha_2 - \beta_2$		-3.8	-3.8					-13.5	-13.5
$\gamma_{\mathrm{E}1}$	-11.3	-5.5	-16.8	-11.3	-5.5	-16.8	-11.4	-5.8	-17.2
$\gamma_{\mathrm{M}1}$	11.3	-1.1	10.2	11.3	3.4	14.7	11.2	3.8	15.0
$\gamma_{\mathrm{E}2}$	11.3	1.1	12.4	11.3	2.6	13.9	11.4	3.0	14.4
$\gamma_{\mathrm{M}2}$	-11.3	1.1	-10.2	-11.3	-0.6	-11.9	-11.2	-0.8	-12.0
$\gamma^{(\mathrm{n})}$		4.4	4.4		0.1	0.1		-0.3	-0.3
$\gamma_\pi^{(\mathrm{n})}$	45.2	4.4	49.6	45.2	12.1	57.3	45.2	13.4	58.6 ± 4.0

These predictions are in strong contradiction to the experimental results except for the backward spin polarizability $\gamma_\pi^{(\mathrm{p})}$, which is dominated by the t-channel π^0 exchange. In ChPT, this is called the Wess–Zumino–Witten term. The difficulties arising in ChPT due to the Δ resonance are strongly correlated with the convergence of the calculation if higher loops are incorporated.

The method used to interpret the experimental cross sections with the help of dispersion relations allows us to pin down some of the parameters within this theoretical approach: (i) the difference $\alpha - \beta$ between the electromagnetic polarizabilities, for which the asymptotic contribution has been modeled by the t-channel exchange of a σ meson; (ii) the backward spin polarizability γ_π, given by the t-channel exchange of a π^0 meson; and (iii) the strengths of the M1 and E2 multipoles involved in the excitation of the Δ-resonance. From the known ingredients, we are now able to predict all the higher-order electromagnetic polarizabilities $\alpha_{\mathrm{E}\nu}$ and $\beta_{\mathrm{M}\nu}$, the quadrupole polarizabilities $\alpha_{\mathrm{E}2}$ and $\beta_{\mathrm{M}2}$, and the four spin polarizabilities $\gamma_{\mathrm{E}1}$, $\gamma_{\mathrm{M}1}$, $\gamma_{\mathrm{E}2}$ and $\gamma_{\mathrm{M}2}$.

In Tables 6.3 and 6.4, the values of the disperion relation approach (DR) described in this book are tabulated in comparison with the predictions of ChPT and of a dispersion relation calculation (HDT) by Drechsel et al. [34]. The HDT calculation used the π multipoles of Hanstein et al. [35] and an upper integration limit of 500 MeV, compared with the 1500 MeV used in the present work. Therefore, only the spin polarizabilities were calculated in the HDT calculation. The reader should keep in mind that no Δ contribution has been given for the ChPT results.

From the discussion above, it becomes clear that the calculations within the ChPT are still not satisfactory. At the one-loop level, the predictions of the higher-order electromagnetic polarizabilities deviate significantly from the predictions of dispersion theory. For example $\beta_{M\nu}$ and β_{M2} deviate by a factor of 2.5! Even the sign of $\alpha_{E\nu}$ does not agree. The inclusion of a Δ contribution shifts the forward and backward spin polarizabilities to comparable values. But then the electromagnetic polarizabilities α and β are in contradiction to the experimental values. It is expected [33] that a calculation within the "small-scale expansion" to the order $\mathcal{O}(\epsilon^4)$ will shed more light on the convergence of the expansion and the contributions of the Δ resonance. It should also be mentioned that in the case of α and β, ChPT does not properly deal with the t-channel exchange of a σ meson, which is an essential degree of freedom of the nucleon for the quantities mentioned above.

References

1. F. J. Federspiel et al., Phys. Rev. Lett. **67** (1991) 1511
2. B. E. MacGibbon et al., Phys. Rev. C **52** (1995) 2097
3. J. Peise et al., Phys. Lett. B **384** (1996) 37
4. A. Hünger et al., Nucl. Phys. A **620** (1997) 385
5. C. Molinari et al., Phys. Lett. B **371** (1996) 181
6. J. Tonnison et al., Phys. Rev. Lett. **80** (1998) 4382
7. G. Blanpied et al., Phys. Rev. Lett. **76** (1996) 1023
8. G. Blanpied et al., Phys. Rev. Lett. **79** (1997) 4337
9. R. Beck et al., Phys. Rev. Lett. **78** (1997) 606
10. R. Beck et al., Phys. Rev. C **61** (2000) 035204
11. M. Camen et al., Phys. Rev. C **65** (2002) 032202
12. A. I. L'vov, A. M. Nathan, Phys. Rev. C **59** (1999) 1064
13. M. Camen, Dissertation, Universität Göttingen (2001), Cuvillier, Göttingen, 2001
14. K. Kossert, Dissertation, Universität Göttingen (2001), Cuvillier, Göttingen, 2001
15. K. Kossert et al., Phys. Rev. Lett. **88** (2002) 162301
16. K. Kossert et al., Eur. Phys. J. A **16** (2003) 259
17. K. W. Rose et al., Nucl. Phys. A **514** (1990) 621
18. N. R. Kolb et al., Phys. Rev. Lett. **85** (2000) 1388
19. R. A. Arndt et al., Phys. Rev. C **53** (1996) 430; the SAID database can be accessed via http://gwdac.phys.gwu.edu
20. D. Drechsel et al., Nucl. Phys. A **645** (1999) 145: the MAID database can be accessed via http://www.kph.uni-mainz.de/MAID/
21. A. I. L'vov, Phys. Lett. B **304** (1993) 29
22. J. Ahrens et al., Phys. Rev. Lett. **84** (2000) 5950
23. J. Ahrens et al., Phys. Rev. Lett. **87** (2001) 022003
24. J. Ahrens et al., Phys. Rev. Lett. **88** (2002) 232002
25. D. Babusci et al., Phys. Rev. C **58** (1998) 1013
26. M. I. Levchuk, A. I. L'vov, Nucl. Phys. A **674** (2000) 449
27. R. Weiner, W. Weise, Phys. Lett. B **159** (1985) 85
28. F. Schöberl, H. Leeb, Phys. Lett. B **166** (1986) 355
29. N. N. Scoccola, W. Weise, Nucl. Phys. A **517** (1990) 495

30. V. Bernard, N. Kaiser, U.-G. Meissner, Nucl. Phys. B **373** (1992) 346
31. V. Bernard, N. Kaiser, U.-G. Meissner, Int. J. Mod. Phys. E **4** (1995) 193
32. T. R. Hemmert, B. R. Holstein, J. Kambor, Phys. Rev. D **55** (1997) 5598
33. T. R. Hemmert et al., Phys. Rev. D **57** (1998) 5746
34. D. Drechsel, G. Krein, O. Hanstein, Phys. Lett. B **420** (1998) 248
35. O. Hanstein, D. Drechsel, L. Tiator, Nucl. Phys. A **632** (1998) 561

7 Summary and Outlook

The experiments on Compton scattering from the proton presented in this book, i.e. the CATS NaI(Tl) experiments [1, 2, 3, 4], the LARA [5] experiment and the TAPS experiments [6], cover an energy range of 55 MeV to 460 MeV and an angular range of $44°$ to $155°$. The electromagnetic polarizabilities α_p and β_p have been extracted from the low-energy TAPS experiment. By including the experiments of Federspiel et al. [7], Zieger et al. [8] and MacGibbon et al. [9], a new global average was determined. If the Baldin sum rule, which has been reevaluated as $\alpha_p + \beta_p = 13.8 \pm 0.4$, is considered, the results are [6]

$$\begin{aligned}
\alpha_p &= 12.1 \pm 0.3(\text{stat.}) \mp 0.4(\text{syst.}) \pm 0.3(\text{mod.}) , \\
\beta_p &= 1.6 \pm 0.4(\text{stat.}) \pm 0.4(\text{syst.}) \pm 0.4(\text{mod.}) .
\end{aligned} \tag{7.1}$$

The uncertainties denote the statistical, systematic and model-dependent uncertainties. The resulting new global average for $\alpha_p - \beta_p$ is

$$\alpha_p - \beta_p = 10.5 \pm 0.9(\text{stat.} + \text{syst.}) \pm 0.7(\text{mod.}) . \tag{7.2}$$

Compared with MacGibbon's global average [9], the statistical precision of the electromagnetic polarizabilities has been improved by a factor of almost 2/3. A detailed investigation of α_p and β_p shows that the asymptotic contribution of the amplitude A_1, which has been modeled by a t-channel 2π exchange, enters with opposite sign into these two quantities. Therefore, this contribution is responsible for the large enhancement of α_p and the large diamagnetic part of β_p. The paramagnetic part of β_p is dominated by the Δ-resonance excitation. Furthermore, it has been shown that α_p, which is related to electric (E1) excitation, has a strong contribution from the Δ resonance which can be attributed to retardation and recoil effects.

When the backward spin polarizability $\gamma_\pi^{(\text{p})}$ was included in the fit procedure as well, no indication was found for an additional contribution to the amplitude A_2. The result obtained from a global analysis of low-energy Compton scattering,

$$\begin{aligned}
\alpha_p &= 12.4 \pm 0.6(\text{stat.}) \mp 0.5(\text{syst.}) \pm 0.1(\text{mod.}) , \\
\beta_p &= 1.4 \pm 0.7(\text{stat.}) \pm 0.4(\text{syst.}) \pm 0.1(\text{mod.}) , \\
\gamma_\pi^{(\text{p})} &= -36.1 \pm 2.1(\text{stat.}) \mp 0.4(\text{syst.}) \pm 0.8(\text{mod.}) ,
\end{aligned} \tag{7.3}$$

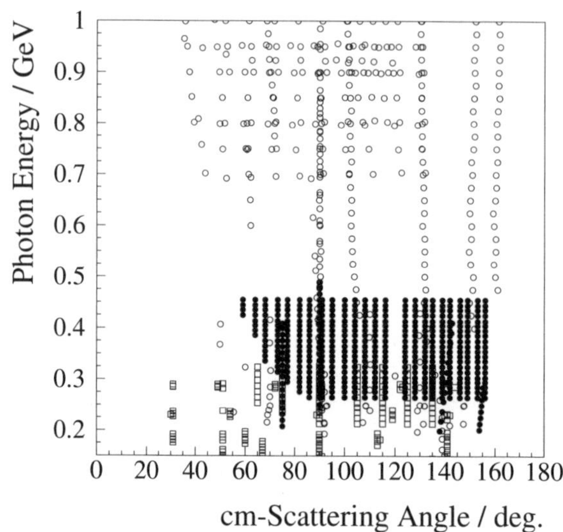

Fig. 7.1. This plot shows a kinematical overview of all experiments performed up to 2000. Only experiments in the Δ-resonance region have been considered. The *filled circles* (●) denote the experiments described in this book. The *open squares* (□) are from other groups. The *open circles* denote all the experiments before 1990 (see Fig. 2.2)

is in good agreement with the findings of the LARA experiment, i.e. Compton scattering above π production threshold, where a value of

$$\gamma_\pi^{(\mathrm{p})} = -37.1 \pm 0.6(\mathrm{stat.} + \mathrm{syst.}) \pm 3.0(\mathrm{mod.}) \, . \tag{7.4}$$

is found (4.23).

All new experimental results lead to the average $\gamma_\pi^{(\mathrm{p})} = -38.7 \pm 1.8$, which is in remarkable contradiction to the value of $\gamma_\pi^{(\mathrm{p})} = -27.1 \pm 2.2^{+2.8}_{-2.4}$ published by the LEGS group [10]. This may be explained by the differences in the measured cross sections at backward angles (see Figs. 4.8 and 4.12). Therefore, no additional contribution to the amplitude A_2 has been found. This statement is in accordance with the new sum rule obtained by L'vov and Nathan, from which $\gamma_\pi^{(\mathrm{p})} = -39.5 \pm 2.4$ was obtained [11].

The available data base for the proton has been extended considerably (Fig. 7.1) in the Δ-resonance region. The interpretation of the differential cross sections in the framework of dispersion relations confirms the validity of this approach [5] when the SAID SM99K π multipoles of Arndt et al. [12] or the π multipoles of MAID2000 [13] are used. The strength of the E2 excitation of the Δ resonance has been determined from the LARA experiment. Expressed in terms of the isospin-3/2 components of the M1 and E2 multipoles, the result

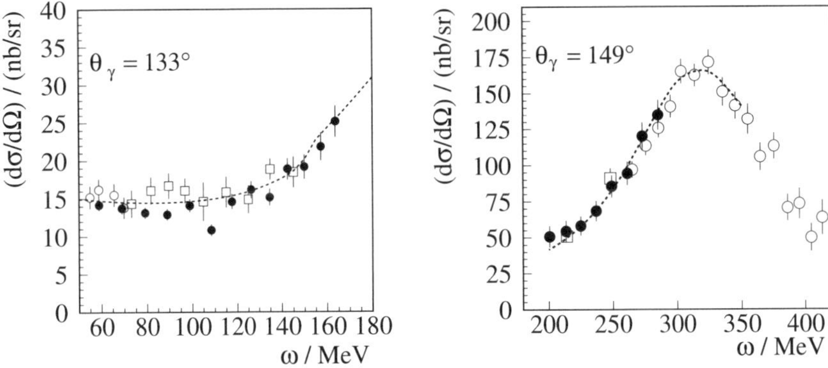

Fig. 7.2. A comparison of the experimental data measured at low energies at $\theta_\gamma = 133°$ (*left*) and in the Δ-resonance region at $\theta_\gamma = 149°$ (*right*). The notation for the data points is the same as in Figs. 4.2 and 5.9. The *dashed line* is the result of Drechsel et al. [22] obtained with subtracted dispersion relations using the π-multipole analysis by Hanstein et al. [23]. The parameters were taken as $\alpha_p + \beta_p = 13.6$, $\alpha_p - \beta_p = 10.7$ and $\gamma_\pi^{(p)} = -37$. The calculation by L'vov et al. [16], which matches the dashed line almost exactly, is not plotted

$$\frac{E2}{M1} = \frac{\operatorname{Im} E_{1+}^{3/2}}{\operatorname{Im} M_{1+}^{3/2}} = (-2.2 \pm 0.3(\text{stat.} + \text{syst.}) \pm 0.2(\text{mod.}))\% \,, \qquad (7.5)$$

is in agreement with π photoproduction experiments by Beck et al. [14, 15] within the uncertainties.

A different approach to describing Compton scattering with fixed-t dispersion relations has been published by Drechsel et al. [22]. In contrast to the approach of L'vov et al. [16], Drechsel et al. used subtracted dispersion relations. The idea is to reduce the model dependence, leaving only two parameters, i.e. $\alpha_p - \beta_p$ and $\gamma_\pi^{(p)}$. The advantage of this method is that the modeling of the amplitudes A_1 and A_2 at large ν can be omitted. The amplitudes converge very rapidly and are saturated by single-π photoproduction at rather moderate energies around 500 MeV. The description of a 2π t-channel exchange was taken from $\gamma\gamma \to \pi\pi$ experiments and thus a detailed modeling of the "σ meson" was avoided. A comparison of the two approaches (Fig. 7.2) does not show a significant difference, neither at low photon energies nor at photon energies in the Δ-resonance region. The only difference is that the model-dependent uncertainty in the extracted values of $\alpha_p - \beta_p$, for example, might be reduced.

The first attempt to investigate quasi-free Compton scattering from the proton bound in the deuteron [3] has been described in this book. The triple differential cross sections, as measured with the TAPS detector from 200 MeV to 300 MeV, could be described by the theoretical calculation of Levchuk et al.

[17]. This excellently verifies that the inclusion of all relevant binding effects is well under control. Using Levchuk's formalism, the difference between the electromagnetic polarizabilities was extracted. The new value of

$$\alpha_{\rm p} - \beta_{\rm p} = 10.0 \pm 1.6 (\text{stat.} + \text{syst.}) \pm 1.1 (\text{mod.}) , \qquad (7.6)$$

determined from the extracted free-proton cross sections using the π photoproduction amplitudes of SAID SM99K and MAID2000, agrees with the new global average obtained from the low-energy TAPS experiment, i.e. $\alpha_{\rm p} - \beta_{\rm p} = 10.5 \pm 0.9 \pm 0.7$. This experiment proved that quasi-free Compton scattering can be separated kinematically from quasi-free π^0 photoproduction in the energy region considered here. The procedure used to extract the free cross sections from the quasi-free results works successfully. The extracted free cross sections are in agreement with the values obtained from the experiments on the free proton. This was a major step forward for the subsequent experiment, which was aimed at Compton scattering from the neutron [18].

It was of great interest to measure quasi-free Compton scattering from the neutron and proton simultaneously and, in addition, Compton scattering from the free proton under the same kinematical conditions. From the experimental point of view, a direct comparison between the free and quasi-free reactions under exactly identical kinematical conditions could then be made. This was the aim of the experiment described in this book which used the CATS NaI(Tl) detector and the neutron detector SENECA [4, 19, 20]. On the basis of the analysis of scattering from a free proton (liquid-hydrogen target), the analysis of the quasi-free scattering from a bound proton (liquid-deuterium target) was optimized in order to achieve the best separation between the scattered events and the background events due to quasi-free π^0 photoproduction. For the first time, Compton scattering from the neutron has been investigated in the entire Δ-resonance region. From the triple differential cross sections at $\theta_\gamma = 136.2°$, the difference between the electromagnetic polarizabilities was determined as follows [19, 20]:

$$\alpha_{\rm n} - \beta_{\rm n} = 9.8 \pm 3.6 (\text{stat.})^{+2.1}_{-1.1} (\text{syst.}) \pm 2.2 (\text{mod.}) . \qquad (7.7)$$

This is a remarkable result, since it confirms for the first time that the electromagnetic polarizabilities of the proton and neutron do not show a significant difference. If the Baldin sum rule for the neutron is included, the electric and magnetic polarizabilities are [19, 20]

$$\alpha_{\rm n} = 12.5 \pm 1.8 (\text{stat.})^{+1.1}_{-0.6} (\text{syst.}) \pm 1.1 (\text{mod.}) ,$$
$$\beta_{\rm n} = 2.7 \mp 1.8 (\text{stat.})^{+0.6}_{-1.1} (\text{syst.}) \mp 1.1 (\text{mod.}) . \qquad (7.8)$$

In agreement with the findings for the proton, the backward spin polarizability $\gamma_\pi^{(\rm n)} = 58.6 \pm 4.0$, as given by the dispersion relation approach, does not have any unknown contribution.

As far as future experiments are concerned, one has to keep in mind that experiments on the proton at far forward angles (see Fig. 7.1) are still needed.

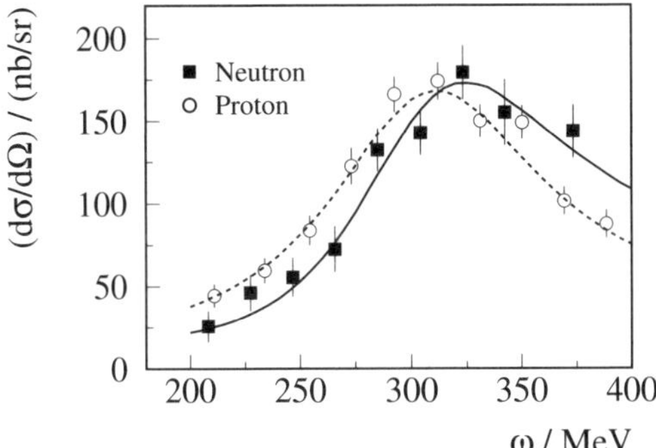

Fig. 7.3. The differential cross sections of the free neutron compared to the proton. The *solid* and *dashed lines* are calculations within the dispersion relation approach using the MAID2000 multipoles

The necessity for new experiments is based on the sensitivity of the differential cross sections to the forward spin polarizability $\gamma^{(\mathrm{p})}$ and the E2/M1 ratio at forward angles. From the experimental point of view, such experiments are rather difficult to perform. The photon detector has to be set up near the photon beam. Owing to the large electromagnetic background produced in the target via pair production, any vetoing system has to be highly segmented in order to reduce the count rate in a single element. The advantage of such an experiment would be that two scattering angles could be covered simultaneously. If the detector is set up at $\theta = 20°$, for example, scattered photons and recoiling protons can be detected at this angle. This means that Compton scattering at $\theta_\gamma = 20°$ (photon detection) and $\theta_\gamma = 130°$ (proton detection) will be measured, since the proton angle of $20°$ corresponds to a photon scattering angle of $\theta_\gamma = 130°$ at $\omega = 250\,\mathrm{MeV}$. This requires a specific detector system in front of the photon detector in order to identify protons. The ratio of the two differential cross sections shows a certain sensitivity to the backward spin polarizability (see Fig. 4.10), provided the electromagnetic polarizabilities α_p and β_p are fixed by the new global average presented in this book. Using the ratio instead of absolute differential cross sections has the advantage that the systematic uncertainties may be reduced, since the normalization to the photon flux and target density can be omitted.

The unpolarized differential cross sections at backward angles are affected by the difference between the electromagnetic polarizabilities $\alpha_\mathrm{p} - \beta_\mathrm{p}$ and by the backward spin polarizability $\gamma_\pi^{(\mathrm{p})}$. At photon energies above the π threshold, the influence of $\gamma_\pi^{(\mathrm{p})}$ is comparable to that of $\alpha_\mathrm{p} - \beta_\mathrm{p}$. Although it has been shown that both quantities can be extracted from unpolarized cross sections measured over a wide energy and angular range, it is important to find an observable which allows one to disentangle the two contributions.

Intuitively one might expect that the polarization degree of freedom would be of great help. For that purpose the following asymmetries [21] for single- and double-polarization experiments could be used for further investigations:

1. The beam asymmetry for linearly polarized photons, where the polarization is either parallel ($\parallel$) or perpendicular ($\perp$) to the scattering plane:

$$\Sigma_3 = \frac{\mathrm{d}\sigma^\parallel - \mathrm{d}\sigma^\perp}{\mathrm{d}\sigma^\parallel + \mathrm{d}\sigma^\perp} \; . \tag{7.9}$$

2. The target asymmetry for unpolarized photons, where the proton is polarized perpendicular ($\pm y$) to the scattering plane:

$$\Sigma_y = \frac{\mathrm{d}\sigma_y - \mathrm{d}\sigma_{-y}}{\mathrm{d}\sigma_y + \mathrm{d}\sigma_{-y}} \; . \tag{7.10}$$

3. The beam–target asymmetry for linearly polarized photons ($\parallel$, $\perp$) and proton polarizations perpendicular ($\pm y$) to the scattering plane:

$$\Sigma_{3y} = \frac{\left(\mathrm{d}\sigma^\parallel - \mathrm{d}\sigma^\perp\right)_y - \left(\mathrm{d}\sigma^\parallel - \mathrm{d}\sigma^\perp\right)_{-y}}{\left(\mathrm{d}\sigma^\parallel + \mathrm{d}\sigma^\perp\right)_y + \left(\mathrm{d}\sigma^\parallel + \mathrm{d}\sigma^\perp\right)_{-y}} \; . \tag{7.11}$$

4. Beam asymmetries for circularly polarized photons, where the polarizations are either right-handed (R) or left-handed (L). The proton spin is aligned in the scattering plane either perpendicular (x) or parallel (z) to the incoming photon momentum:

$$\Sigma_{2x} = \frac{\mathrm{d}\sigma_x^{\mathrm{R}} - \mathrm{d}\sigma_x^{\mathrm{L}}}{\mathrm{d}\sigma_x^{\mathrm{R}} - \mathrm{d}\sigma_x^{\mathrm{L}}} \; , \tag{7.12}$$

$$\Sigma_{2z} = \frac{\mathrm{d}\sigma_z^{\mathrm{R}} - \mathrm{d}\sigma_z^{\mathrm{L}}}{\mathrm{d}\sigma_z^{\mathrm{R}} - \mathrm{d}\sigma_z^{\mathrm{L}}} \; . \tag{7.13}$$

These asymmetries and the unpolarized differential cross section at a scattering angle $\theta_\gamma = 90°$ are shown in Fig. 7.4 as a function of the incident photon energy for various choices of $\gamma_\pi^{(\mathrm{p})}$. The calculations were done with the dispersion relation approach, as used elsewhere in the present book. The huge effect at energies above 350 MeV can be partially compensated by changing the σ mass parameter. Therefore, only the energy region below 350 MeV is of further interest. The various values for $\gamma_\pi^{(\mathrm{p})}$ show the largest effect in Σ_{2x} around 230 MeV. The angular distributions of these quantities at 230 MeV (Fig. 7.5) show that at far backward angles, the asymmetry Σ_{2z} is also dependent on $\gamma_\pi^{(\mathrm{p})}$. It follows from the calculated asymmetries that Σ_{2x} at $\theta_\gamma \approx 120°$ seems to be best suited for experimental investigations. However, the differences are rather small if expressed in absolute numbers. For $\gamma_\pi^{(\mathrm{p})} = -27, -37, -47$, one obtains $\Sigma_{2x} = -0.49, -0.45, -0.35$, respectively. From the experimental point of view, such a precision is hard to achieve.

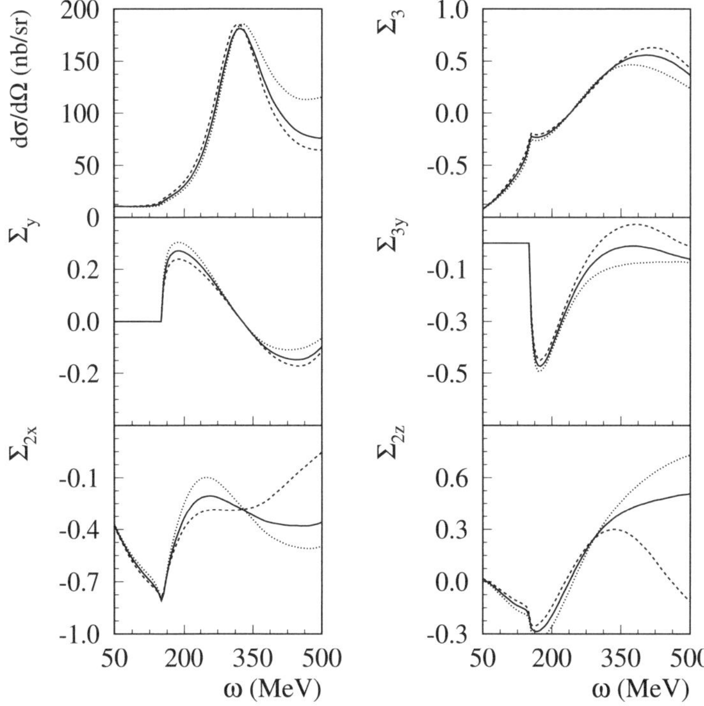

Fig. 7.4. The unpolarized differential cross section (*top left*) and the asymmetries Σ_3, Σ_y, Σ_{3y}, Σ_{2x} and Σ_{2z} as a function of the incident photon energy at $\theta_\gamma = 90°$ (laboratory). The calculations were performed using $\gamma_\pi^{(P)} = -37$ (*solid lines*), $\gamma_\pi^{(P)} = -27$ (*dashed lines*) and $\gamma_\pi^{(P)} = -47$ (*dotted lines*)

The polarized differential cross sections themselves exhibit a strong dependence on $\gamma_\pi^{(P)}$ and a moderate dependence on $\alpha_p - \beta_p$. Figure 7.6 shows the polarized differential cross section at $\omega = 230$ MeV for circularly polarized photons with the proton spin aligned in the scattering plane perpendicular to the momentum of the incoming photon. The dependence on $\gamma_\pi^{(P)}$ is plotted in the left column. The influence of $\alpha_p - \beta_p$ is given in the right column. In Fig. 7.7, corresponding plots are shown for the proton spin aligned parallel and antiparallel to the momentum of the incident photon. The influence of $\alpha_p - \beta_p = 11 \pm 2$, as shown in Figs. 7.6 and 7.7, is enhanced by a factor 2. The new global average gives a total uncertainty of about ± 1. Thus, the influence of $\alpha_p - \beta_p$ may be neglected. The conclusion is that a measurement of the polarized cross section for circularly polarized photons and polarized protons, with the proton spin aligned in the scattering plane, at $\theta_\gamma = 135°$ and at incident photon energies between 200 MeV and 300 MeV, provides a possibility to extract the backward spin polarizability $\gamma_\pi^{(P)}$.

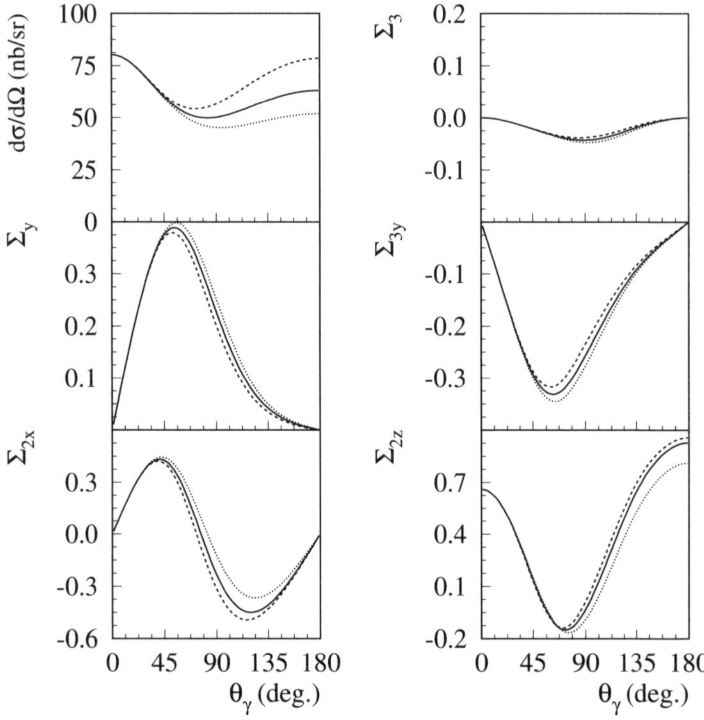

Fig. 7.5. The unpolarized differential cross section (*top left*) and the asymmetries Σ_3, Σ_y, Σ_{3y}, Σ_{2x} and Σ_{2z} as a function of the scattering angle (laboratory) at $\omega = 230$ MeV. The calculations were performed using $\gamma_\pi^{(\mathrm{P})} = -37$ (*solid lines*), $\gamma_\pi^{(\mathrm{P})} = -27$ (*dashed lines*) and $\gamma_\pi^{(\mathrm{P})} = -47$ (*dotted lines*)

Using circularly polarized photons, and protons polarized parallel to the momentum of the incoming photon (γ–p helicity state 3/2) or antiparallel to it (γ–p helicity state 1/2), gives direct access to the amplitude A_4 at forward scattering angles. The imaginary part of this amplitude determines the value associated with the Gerasimov–Drell–Hearn sum rule (see (3.42) and (3.75)) and the forward spin polarizability $\gamma^{(\mathrm{P})}$. The validity of the GDH sum rule has been questioned [24, 25], since the prediction based on the partial-wave analysis of Arndt et al. [12] is incompatible with the sum rule. A possible violation of the GDH sum rule might be explained by the high-energy behavior of the real part of A_4, i.e. the asymptotic contribution A_4^{as}. In the dispersion formalism used in this book it was assumed that A_4^{as} is small enough to be neglected. The truth may be found by double-polarized experiments. Experimental attempts have been started to measure the helicity-dependent absorption cross section of the proton up to 3 GeV at MAMI (Mainz) [26, 27, 28] and at ELSA (Bonn). The first results obtained at MAMI [26, 27] indicate that the GDH

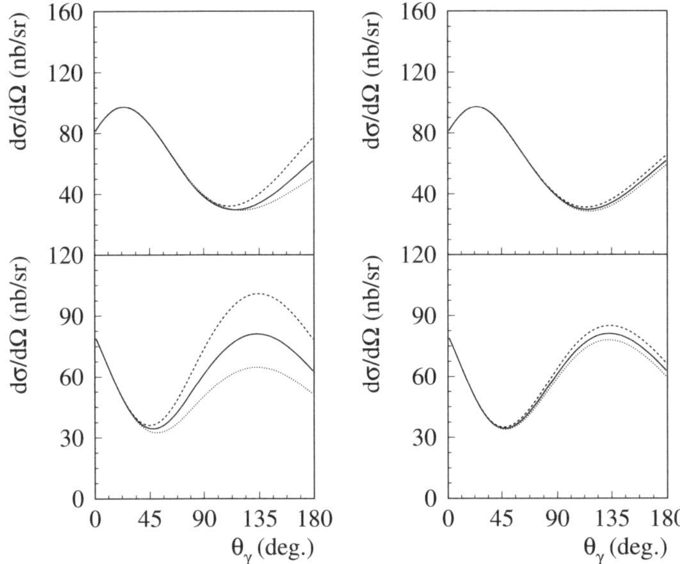

Fig. 7.6. The polarized differential cross section at $\omega = 230$ MeV for circularly polarized photons with the proton spin aligned in the scattering plane perpendicular to the momentum of the incident photon. *Top*: the proton spin is aligned along the positive x direction. *Bottom*: the proton spin is aligned along the negative x direction. *Left column*: the calculations use $\gamma_\pi^{(\mathrm{p})} = -37$ (*solid lines*), $\gamma_\pi^{(\mathrm{p})} = -27$ (*dashed lines*) and $\gamma_\pi^{(\mathrm{p})} = -47$ (*dotted lines*). *Right column*: the calculations use $\alpha_\mathrm{p} - \beta_\mathrm{p} = 11$ (*solid lines*), $\alpha_\mathrm{p} - \beta_\mathrm{p} = 9$ (*dashed lines*) and $\alpha_\mathrm{p} - \beta_\mathrm{p} = 13$ (*dotted lines*)

sum rule (see (3.42)) is not violated as previously predicted in many theoretical publications (see [27]). Because of the limited photon energy range of these experiments (200 MeV to 800 MeV), a much better result may obtained for the forward spin polarizability, since at 800 MeV the corresponding sum rule (see (3.44)) is almost saturated. Thus, it is mainly the contribution from the energy range below 200 MeV that needs to be introduced by means of a theoretical estimate. The final result is in good agreement with the results obtained from the analysis of Compton scattering using the dispersion relation approach.

Future Compton scattering experiments will clearly be driven by the need to use polarized photon beams and polarized targets. The production of linearly and circularly polarized photon beams is already standard at various laboratories, for example MAMI (Mainz, Germany) [26, 29] and LEGS (Brookhaven, USA) [30]. The experimental difficulties arising are mainly related to the polarized targets. The ideal target would, of course, be liquid hydrogen. But at low temperatures molecular hydrogen has a total spin of

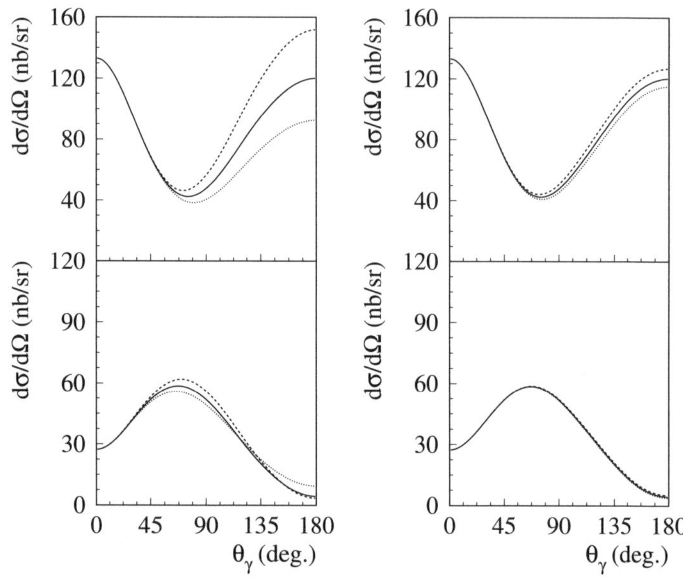

Fig. 7.7. The polarized differential cross section at $\omega = 230$ MeV for circularly polarized photons with the proton polarization parallel (*top*) and antiparallel (*bottom*) to the momentum of the incident photon. The notation is the same as in Fig. 7.6

zero (para-hydrogen) and is not polarizable at all. Therefore, other materials have to be used which have a sufficient content of polarizable protons [31]. Typical materials are butanol (C_4H_9OH) and ammonia (NH_3). Both targets have the disadvantage that, in addition to hydrogen, the molecules contain a large fraction of carbon and oxygen (butanol) or nitrogen (ammonia). Therefore, Compton scattering from the free proton is accompanied by the following reactions:

1. Elastic and inelastic photon scattering from ^{12}C, ^{16}O or ^{14}N.
2. Incoherent photon scattering from ^{12}C, ^{16}O or ^{14}N.
3. Coherent π^0 photoproduction from ^{12}C, ^{16}O or ^{14}N.
4. Incoherent π^0 photoproduction from ^{12}C, ^{16}O or ^{14}N.

All these additional reactions lead to photons in the final state and are the origin of background in the spectra of photons elastically scattered from the free protons. Photons scattered from nuclei heavier than the proton can be identified by the scattered photon energy. For example, at $\omega = 230$ MeV and $\theta_\gamma = 135°$ the scattered photon energies are 162 MeV and 222 MeV for the proton and the carbon nucleus, respectively. These photons can be efficiently suppressed if the scattered photons and the recoiling protons are detected in coincidence. The same holds for coherent π^0 photoproduction from ^{12}C, ^{16}O or ^{14}N. For such photon–proton coincidence measurements, the following incoherent reactions remain as the only source of background which has to be subtracted:

1. It is difficult to separate incoherent scattering from ^{12}C, ^{16}O or ^{14}N from the free-proton scattering. The energy resolution of the photon and proton detectors may help because the separation energy of the bound proton is about 16 MeV for these nuclei. In any case, however, to subtract this kind of background, very precise measurements on pure carbon, oxygen or nitrogen targets are required.
2. Incoherent π^0 photoproduction from ^{12}C, ^{16}O or ^{14}N has to be subtracted. Thus, additional measurements on pure carbon, oxygen or nitrogen targets are required.

The discussion above shows that double-polarized Compton scattering experiments would require an enormous effort. For technical reasons, the polarized target to be installed is much more complex than a liquid-hydrogen target. Experimentally, the beam time consumption would exceed the 100 h to 200 h which is sufficient for unpolarized experiments. Additional experiments on carbon, oxygen or nitrogen with high statistical precision would have to be done. It may be questioned whether the precision that can be reached in double-polarized experiments is adequate for improving on the information gained from very precise unpolarized experiments. This question has to be answered by detailed theoretical investigations and by intense simulations of possible detector systems. Therefore, double-polarized Compton scattering experiments will a very challenging task, both experimentally and from the physics point of view.

References

1. J. Peise et al., Phys. Lett. B **384** (1996) 37
2. A. Hünger et al., Nucl. Phys. A **620** (1997) 385
3. F. Wissmann et al., Nucl. Phys. A **660** (1999) 232
4. M. Camen et al., Phys. Rev. C **65** (2002) 032202
5. S. Wolf et al., Eur. Phys. J. A **12** (2001) 231
6. V. Olmos de León et al., Eur. Phys. J. A **10** (2001) 207
7. F. J. Federspiel et al., Phys. Rev. Lett. **67** (1991) 1511
8. A. Zieger et al., Phys. Lett. B **278** (1992) 34
9. B. E. MacGibbon et al., Phys. Rev. C **52** (1995) 2097
10. J. Tonnison et al., Phys. Rev. Lett. **80** (1998) 4382
11. A. I. L'vov, A. M. Nathan, Phys. Rev. C **59** (1999) 1064
12. R. A. Arndt et al., Phys. Rev. C **53** (1996) 430; the SAID database can be accessed via `http://gwdac.phys.gwu.edu`
13. D. Drechsel et al., Nucl. Phys. A **645** (1999) 145; the MAID database can be accessed via `http://www.kph.uni-mainz.de/MAID/`
14. R. Beck et al., Phys. Rev. Lett. **78** (1997) 606
15. R. Beck et al., Phys. Rev. C **61** (2000) 035204

16. A. I. L'vov, V. A. Petrun'kin, M. Schumacher, Phys. Rev. C **55** (1997) 359
17. M. I. Levchuk, A. I. L'vov, V.A. Petrun'kin, Preprint 86, FIAN, Moscow, 1986; Few-Body Syst. **16** (1994) 101
18. F. Wissmann, Proposal A2-9/97, MAMI, Mainz, 1997
19. K. Kossert et al., Phys. Rev. Lett. **88** (2002) 162301
20. K. Kossert et al., Eur. Phys. J. A **16** (2003) 259
21. D. Babusci et al., Phys. Rev. C **58** (1998) 1013
22. D. Drechsel et al., Phys. Rev. C **61** (1999) 015204; B. Pasquini, Institut für Kernphysik, Universität Mainz, private communication
23. O. Hanstein, D. Drechsel, L. Tiator, Nucl. Phys. A **632** (1998) 561
24. A. M. Sandorfi, C. S. Whisnant, M. Khandaker, Phys. Rev. D **50** (1994) R6681
25. S. D. Bass, Mod. Phys. Lett. A **12** (1997) 1051
26. J. Ahrens et al., Phys. Rev. Lett. **84** (2000) 5950
27. J. Ahrens et al., Phys. Rev. Lett. **87** (2001) 022003
28. J. Ahrens et al., Phys. Rev. Lett. **88** (2002) 232002
29. F. Rambo et al., Phys. Rev. C **58** (1998) 489
30. C. E. Thron et al., Nucl. Instrum. Methods A **285** (1989) 447
31. D. G. Crabb, W. Meyer, Ann. Rev. Nucl. Part. Sci. **47** (1997) 67

A Appendix

A.1 Units

If not stated otherwise in the text, the following units have been used throughout this book:

mass, energy	m, ω	MeV
fine-structure constant	$e^2, e^2/4\pi$	1/137.036
conversion constant	$\hbar c$	197.327 MeV fm
classical proton radius	e^2/m	1.53×10^{-3} fm
magnetic moment	μ	$\mu_{\mathrm{N}} = e/2m_{\mathrm{p}}$
scattering amplitudes	f	fm
scattering angle	z	$cos\,\theta_\gamma$
electromagnetic polarizabilities	α, β	10^{-4} fm^3
spin polarizabilities	γ, γ_π	10^{-4} fm^4
	$\gamma_{\mathrm{E1}}, \gamma_{\mathrm{M1}}$	10^{-4} fm^4
	$\gamma_{\mathrm{E2}}, \gamma_{\mathrm{M2}}$	10^{-4} fm^4
dispersion polarizabilities	$\alpha_{\mathrm{E}\nu}, \beta_{\mathrm{M}\nu}$	10^{-4} fm^5
quadrupole polarizabilities	$\alpha_{\mathrm{E2}}, \beta_{\mathrm{M2}}$	10^{-4} fm^5

A.2 Photoabsorption and π Photoproduction Multipoles

The decomposition of the photoabsorption cross section into partial reaction channels permits the investigation of the π photoproduction multipoles
involved in these channels. In particular, for single-π photoproduction, the
amplitudes were defined by Chew, Goldberger, Low and Nambu [1] and are
commonly known as the CGLN amplitudes In the π–nucleon cm frame, the
differential cross section for single-π photoproduction is written as [2]

$$\frac{\mathrm{d}\sigma}{\mathrm{d}\Omega} = \frac{q}{k} \sum |T_{fi}|^2 \,, \tag{A.1}$$

where the summation sign represents summing over the photon polarizations and summing and averaging over the nucleon states, q is the momentum of the emitted π meson, and $k = |\boldsymbol{k}| = \omega$ is the momentum corresponding to the incident photon energy. In the CGLN notation, the T-matrix reads as

$$
T_{fi} = \left\langle f \left| i\boldsymbol{\sigma} \cdot \boldsymbol{\epsilon} F_1 + \frac{(\boldsymbol{\sigma} \cdot \boldsymbol{q})(\boldsymbol{\sigma} \cdot (\boldsymbol{k} \times \boldsymbol{\epsilon}))}{qk} F_2 \right.\right.
$$
$$
\left.\left. + i\frac{(\boldsymbol{\sigma} \cdot \boldsymbol{k})(\boldsymbol{q} \cdot \boldsymbol{\epsilon})}{qk} F_3 + i\frac{(\boldsymbol{\sigma} \cdot \boldsymbol{q})(\boldsymbol{q} \cdot \boldsymbol{\epsilon})}{q^2} F_4 \right| i \right\rangle , \tag{A.2}
$$

where $\boldsymbol{\sigma}$ is the Pauli spinor, $\boldsymbol{\epsilon}$ is the photon polarization vector, $\boldsymbol{k}$ is the photon momentum and $\boldsymbol{q}$ is the π meson momentum. The connection between (A.2) and the multipole representation, i.e. the relation between the CGLN amplitudes F_i and the multipole amplitudes $E_{l\pm}$ and $M_{l\pm}$, is given by

$$
F_1 = \sum_{l=0}^{N} (lM_{l+} + E_{l+})P'_{l+1}(z) + ([l+1]M_{l-} + E_{l-})P'_{l-1}(z) , \tag{A.3}
$$

$$
F_2 = \sum_{l=0}^{N} ([l+1]M_{l+} + lM_{l-})P'_l(z) , \tag{A.4}
$$

$$
F_3 = \sum_{l=0}^{N} (E_{l+} - M_{l+})P''_{l+1}(z) + (E_{l-} + M_{l-})P''_{l-1}(z) , \tag{A.5}
$$

$$
F_4 = \sum_{l=0}^{N} (M_{l+} - E_{l+} - M_{l-} - E_{l-})P''_l(z) . \tag{A.6}
$$

Here, the functions P are the derivatives of the Legendre polynomials and z is the cosine of the π emission angle in the π–nucleon cm frame. Turning to the helicity amplitudes, Walker [3] has established a connection between the CGLN amplitudes and the absorption cross sections of definite helicity. The decomposition of the helicity amplitudes into multipole amplitudes [4] yields the following expressions for the photoabsorption cross sections with helicity 1/2 and 3/2 [5]:

$$
\sigma_{1/2} = \frac{8\pi q}{k} \sum_{n=0}^{5} (n+1)\left(|A_{n+}|^2 + |A_{(n+1)-}|^2 \right) , \tag{A.7}
$$

$$
\sigma_{3/2} = \frac{8\pi q}{k} \sum_{n=0}^{5} \frac{1}{4}n(n+1)(n+2)\left(|B_{n+}|^2 + |B_{(n+1)-}|^2 \right) . \tag{A.8}
$$

The helicity amplitudes are expressed in terms of the multipole amplitudes as

$$
A_{n+} = \frac{1}{2}\left([n+2]E_{n+} + nM_{n+} \right) , \tag{A.9}
$$

$$A_{(n+1)-} = \frac{1}{2}\left([n+2]M_{(n+1)-} - nE_{(n+1)-}\right) , \tag{A.10}$$

$$B_{n+} = E_{n+} - M_{n+} , \tag{A.11}$$

$$B_{(n+1)-} = E_{(n+1)-} + M_{(n+1)-} . \tag{A.12}$$

The restriction within the summation in (A.7) and (A.8) to $N = 5$ reflects the assumption that up to 1700 MeV photon energy the single-π photoproduction is completely described by the multipoles up to $l = 5$. With (A.7) and (A.8), we immediately obtain the total photoabsorption cross section:

$$\sigma_{\text{tot}} = \frac{1}{2}\left(\sigma_{1/2} + \sigma_{3/2}\right) . \tag{A.13}$$

A.3 Relations Between the Invariant Amplitudes

In Sect. 3.2, the definitions of the invariant amplitudes T_i given by Prange [6] and of A_i given by L'vov [7, 8] were introduced. Unfortunately, these are not commonly accepted definitions. The T_i and the amplitudes A_i^{HL} used by Hearn and Leader [9] have the following relations [10]:

$$\begin{aligned}
T_1 &= A_1^{\text{HL}} , & T_4 &= -A_5^{\text{HL}} , \\
T_2 &= -A_4^{\text{HL}} , & T_5 &= -A_3^{\text{HL}} , \\
T_3 &= A_2^{\text{HL}} , & T_6 &= -A_6^{\text{HL}} .
\end{aligned} \tag{A.14}$$

The amplitudes A_i and the similar amplitudes A_i^{BT} introduced by Bardeen and Tung [11] are transformed into each other in the following way [8]:

$$\begin{aligned}
A_1 &= -\frac{1}{2}A_1^{\text{BT}} , \\
A_2 &= \frac{m}{2}A_2^{\text{BT}} , \\
A_3 &= -\frac{m^2}{4}A_5^{\text{BT}} - \frac{m^2}{4\nu}A_6^{\text{BT}} , \\
A_4 &= -\frac{m^2}{4\nu}A_6^{\text{BT}} , \\
A_5 &= \frac{m}{2\nu}A_3^{\text{BT}} , \\
A_6 &= -\frac{m}{2}A_4^{\text{BT}} + \frac{4m^2 - t}{16\nu}A_6^{\text{BT}} .
\end{aligned} \tag{A.15}$$

To express the invariant amplitudes A_i in terms of the helicity amplitudes $T_{\lambda_\gamma' \lambda_N' \lambda_\gamma \lambda_N}$ introduced in Sect.3.2.2 the reduced helicity amplitudes τ_i defined in (3.64) can be used. The relations between A_i and τ_i are the following [8]:

$$A_1 = \frac{1}{(s-m^2)^2}\left[-\frac{s}{m}\left(1-\sigma\frac{s+m^2}{2s}\right)\tau_4 - \frac{\sqrt{s}}{2}(\tau_5+\sigma\tau_6)\right] , \qquad (A.16)$$

$$A_2 = \frac{1}{(s-m^2)^3}\left[-\frac{s}{m}(s+m^2)\left(1-\sigma\frac{s-m^2}{2s}\right)\tau_4\right.$$
$$\left.-\frac{\sqrt{s}}{2}(s-m^2)\tau_5 + 2s\sqrt{s}\left(1-\sigma\frac{s-m^2}{4s}\right)\tau_6\right] , \qquad (A.17)$$

$$A_3 = \frac{1}{(s-m^2)^2(s-m^2+t/2)}\left[m^3[\tau_1+(1-\sigma)\tau_2]\right.$$
$$\left.-2m^2\sqrt{s}\left(1-\sigma\frac{s+m^2}{2s}\right)\tau_3\right] , \qquad (A.18)$$

$$A_4 = \frac{1}{(s-m^2)^2(s-m^2+t/2)}\left[m^3\tau_1 - m^3\left(1+\sigma\frac{m^2}{s}\right)\tau_2\right.$$
$$\left.+\frac{2m^4}{\sqrt{s}}\sigma\tau_3\right] , \qquad (A.19)$$

$$A_5 = \frac{1}{(s-m^2)^2(s-m^2+t/2)}\left[m(s+m^2)\sigma\tau_4\right.$$
$$\left.-m^2\sqrt{s}(\tau_5+\sigma\tau_6)\right] , \qquad (A.20)$$

$$A_6 = \frac{1}{(s-m^2)^2(s-m^2+t/2)}\left[-\frac{m}{2}(s+m^2)(\tau_1+(1-\sigma)\tau_2)\right.$$
$$\left.+2m^2\sqrt{s}(1-\sigma)\tau_3\right] . \qquad (A.21)$$

Here, s and t are the usual Mandelstam variables and

$$\sigma = \sin^2\frac{\theta_\gamma}{2} = -\frac{st}{(s-m^2)^2} , \qquad (A.22)$$

where θ_γ is the photon scattering angle in the cm system.

To calculate the differential cross section for Compton scattering, one may transform the amplitudes A_i into the amplitudes T_i which define the total scattering amplitude, as given in (3.49):

$$T_1 = \frac{t}{2}(A_1+A_6) + \frac{m^4-su}{2m^2}A_3 - 2\nu^2 A_5 ,$$
$$T_2 = 2\nu(A_5+A_6) ,$$
$$T_3 = \frac{t}{2}(A_1-A_6) - \frac{m^4-su}{m^2}A_3 - 2\nu^2 A_5 ,$$
$$T_4 = 2\nu(A_5-A_6) ,$$
$$T_5 = \frac{t}{2}A_2 - 2\nu^2 A_5 ,$$
$$T_6 = \frac{m^4-su}{2m^3}A_4 + \frac{t}{2m}A_6 . \qquad (A.23)$$

A.4 Tagged Photon Beam at MAMI

The tagged bremsstrahlung beam [12, 13] installed at the electron accelerator MAMI in Mainz [14, 15] was used for all experiments described in the present book. MAMI is a cascaded cw racetrack microtron (RTM) (Fig. A.1). The electron gun injects electrons into the linear accelerator (linac) stage, which accelerates the electrons to an energy of 3.56 MeV. The linac serves as an injector for the following RTM stages. After the linac, the electrons are accelerated in three subsequent stages RTM1, RTM2 and RTM3 to energies of 14.4 MeV, 179.8 MeV and 855 MeV, respectively. MAMI can deliver a maximum beam intensity of approximately 100 μA. However, for experiments such as those presented here the electron beam intensity is limited by the tagging system and is typically lower than 100 nA.

The system that creates the tagged photon beam is installed in the Tagger Hall. The main parts of the tagging system are outlined in Fig A.2. The MAMI electron beam hits a radiator, typically a 4 μm thin Ni foil, in front of the tagging spectrometer and is then momentum analyzed with a dipole magnet. The bremsstrahlung photons produced in the radiator are emitted in the direction of the incident electrons in a rather narrow cone. The photon beam then passes through a collimation system within the yoke of the magnet about 2.5 m downstream of the radiator.

Electrons which do not emit bremsstrahlung photons are deflected by 80°, led out of the experimental hall and dumped in a Faraday cup. Electrons which emit bremsstrahlung photons with energies of interest are detected in the spectrometer focal plane by 352 overlapping plastic scintillators. A single tagging channel is defined by the coincidence of two detectors. Thus, a total of 351 tagging channels are available. The acceptance of the tagging system ranges from 5% to 94% of the incident electron momentum. This means that the tagged-photon energies range between 50 MeV and 800 MeV, with an energy width of the tagging channels between 2.5 MeV and 1.5 MeV, respectively. The tagged-photon flux is limited by the counting rate of the tagging channel corresponding to the lowest photon energy and is commonly chosen to be 0.5 MHz. By switching off the tagging channels at lower photon energies, the number of tagged photons can be optimized according to the experimental requirements.

The collimation of the photon beam has to be chosen according to the desired beam spot size on the target and depends on the distance of the target from the radiator. The collimation also defines the tagging efficiency, which is necessary to determine the photon flux on the target. The tagging efficiency $\varepsilon_{\mathrm{tag}}$ is given by the ratio of tagged photons in the photon beam N_γ to the number of electrons N_{e} detected by the focal-plane detectors:

$$\varepsilon_{\mathrm{tag}} = \frac{N_\gamma}{N_{\mathrm{e}}}. \tag{A.24}$$

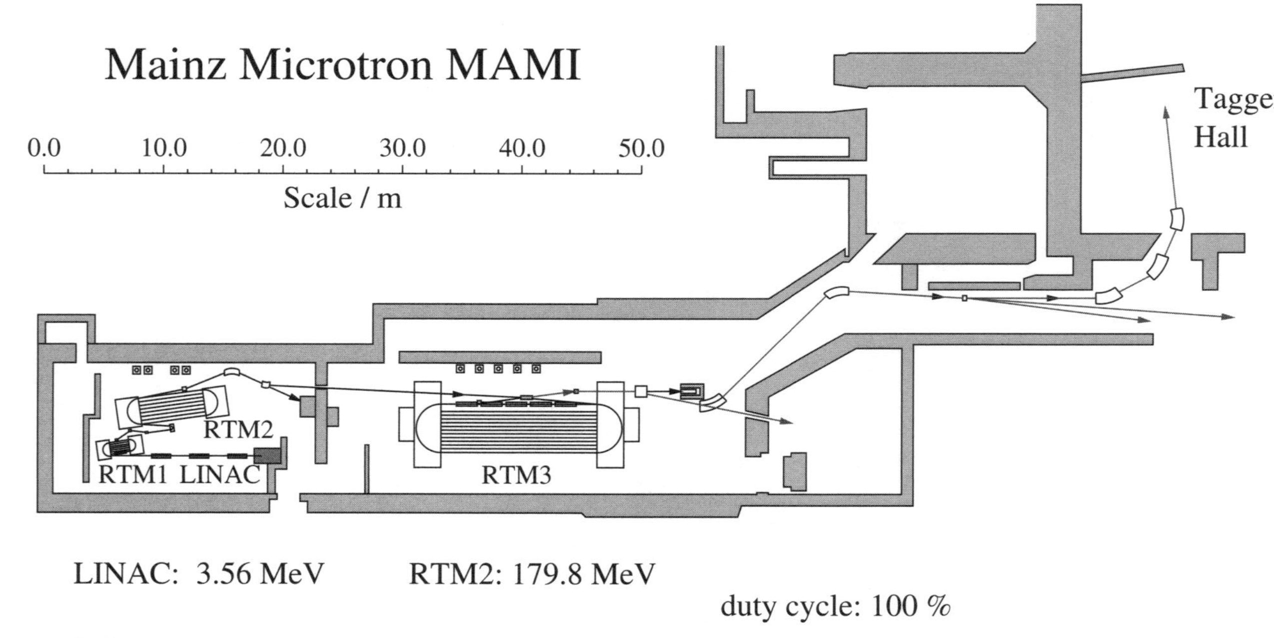

Fig. A.1. Floor plan of the **MA**inz **MI**crotron (MAMI). The electrons are accelerated after the linac by a three-stage microtron (RTM1, RTM2, RTM3) to an end energy of 855 MeV maximum. After RTM3, the beam transfer system distributes the electron beam to the various experimental halls. The experiments presented in this book were carried out in the Tagger Hall. With permission from Institut für Kernphysik, Universität Mainz

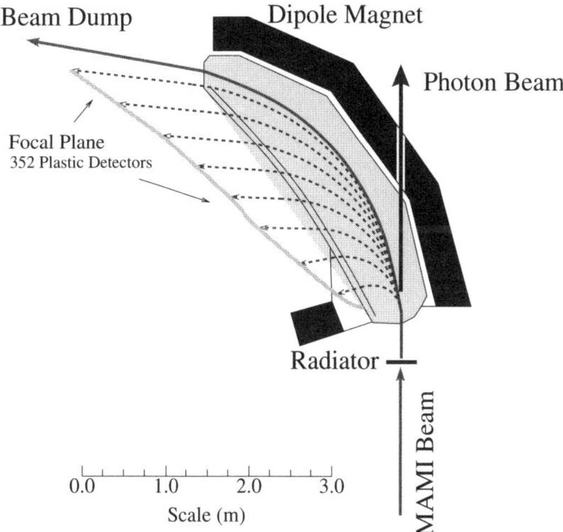

Fig. A.2. The Glasgow–Mainz tagging system at MAMI. With permission from Institut für Kernphysik, Universität Mainz

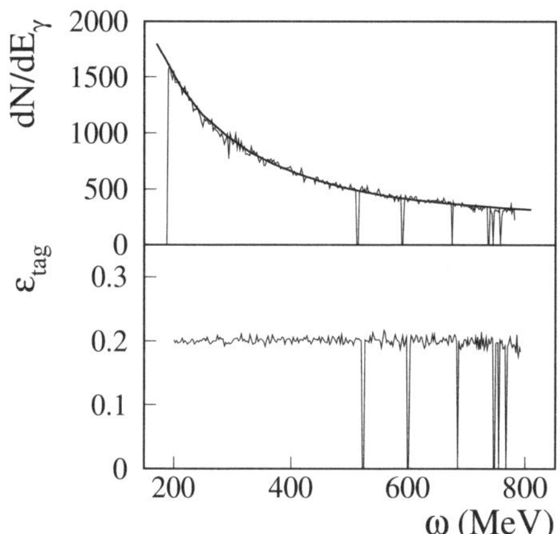

Fig. A.3. *Top*: tagger electron spectrum, as measured in coincidence with the Pb glass detector in the direct photon beam. The *solid line* is the result of a theoretical calculation [16]. *Bottom*: tagging efficiency as measured with the Pb glass detector. Both spectra are examples obtained during the scattering experiment with the CATS NaI(Tl) detector at $\theta_\gamma = 130.7°$. The rather low tagging efficiency of about 20% is the result of a small photon beam collimator behind the tagger

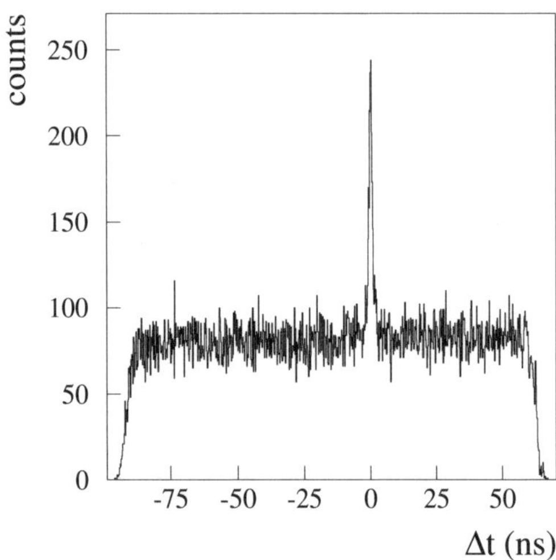

Fig. A.4. Example of the measured time differences between the CATS NaI(Tl) detector and the tagger during a Compton scattering experiment. The coincidence peak is located on a flat background of random coincidences. The time resolution given by the full-width half-maximum of the peak is about 1 ns

Throughout an experiment, this quantity needs to be measured several times with a totally absorbing Pb glass or BGO detector in the photon beam. Between two tagging efficiency measurements, ε_{tag} is monitored by a P2-type ionization chamber in the photon beam dump at the end of the beam line. Figure A.3 shows a typical result of a tagging efficiency measurement.

During an experiment, only those incident photons are tagged which interact with the target so as to lead to reaction products in the detectors which create the trigger signal. The time difference between the trigger signal and the corresponding tagging channel serves as the coincidence signal. The correlated events show up as a sharp peak in such a time spectrum (Fig. A.4). Because of the huge count rate of the tagger (≈ 100 MHz), there are random coincidences as well. With a restrictive time cut around the coincidence peak, most of the random events can be suppressed. The remaining background has to be subtracted by doing the same analysis with a time cut outside the coincidence peak.

A.5 Kinematical Description of Compton Scattering

The calculation of the kinematical quantities is based on energy and momentum conservation. For a two-body process such as Compton scattering, the conservation laws have the form

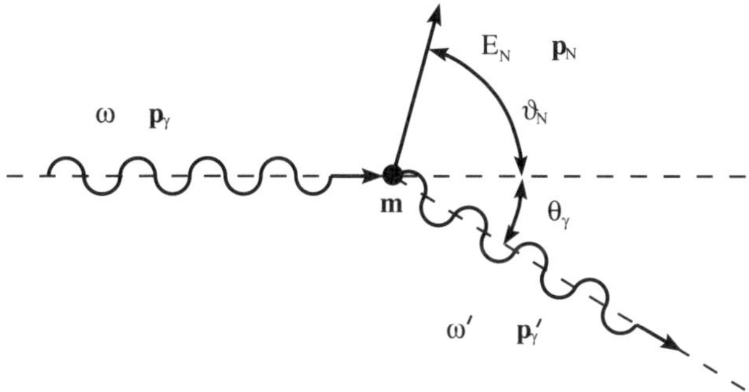

Fig. A.5. Kinematics of Compton scattering. ω, ω' and $\boldsymbol{p}_\gamma$, $\boldsymbol{p}_\gamma'$ are the energy and the momentum of the incident and the scattered photon, respectively. $E_{\rm N}$, $\boldsymbol{p}_{\rm N}$ are the total energy and momentum of the recoiling nucleon, and θ_γ and $\vartheta_{\rm N}$ are the corresponding emission angles

$$\omega + m = \omega' + E_{\rm N} , \qquad \boldsymbol{p}_\gamma = \boldsymbol{p}_\gamma' + \boldsymbol{p}_{\rm N} . \tag{A.25}$$

Here, ω and ω' are the energies of the incident and the scattered photon, respectively, $\boldsymbol{p}_\gamma$ and $\boldsymbol{p}_\gamma'$ are their momenta, m denotes the mass of the struck nucleon N, and $E_{\rm N}$ and $\boldsymbol{p}_{\rm N}$ are its total energy and its momentum (Fig. A.5). The energy of the scattered photon is calculated by eliminating the quantities related to the nucleon in (A.25). The well-known result is

$$\omega' = \frac{\omega}{1 + (\omega/m)\left(1 - \cos\theta_\gamma\right)} , \tag{A.26}$$

where θ_γ is the photon scattering angle (laboratory system).

Likewise, the kinematical quantities of the scattered photon may be eliminated, and we obtain the total energy of the recoiling nucleon,

$$E_{\rm N} = m\frac{\omega^2\left(1 + \cos^2\vartheta_{\rm N}\right) + m(m + 2\omega)}{\omega^2(1 - \cos^2\vartheta_{\rm N}) + m(m + 2\omega)} , \tag{A.27}$$

where $\vartheta_{\rm N}$ is the recoil angle of the nucleon (laboratory system).

Using momentum conservation, a relation between the momentum components perpendicular ($\perp$) and parallel ($\|$) to the momentum of the incident photon can be deduced:

$$p_{\gamma\|}' = p_\gamma' \cos\theta_\gamma , \qquad p_{\rm N\|} = p_{\rm N} \cos\vartheta_{\rm N} , \tag{A.28}$$

$$p_{\gamma\perp}' = p_\gamma' \sin\theta_\gamma , \qquad p_{\rm N\perp} = p_{\rm N} \sin\vartheta_{\rm N} , \tag{A.29}$$

$$p_\gamma = p_{\gamma\|}' + p_{\rm N\|} , \tag{A.30}$$

$$p_{\gamma\perp}' + p_{\rm N\perp} = 0 . \tag{A.31}$$

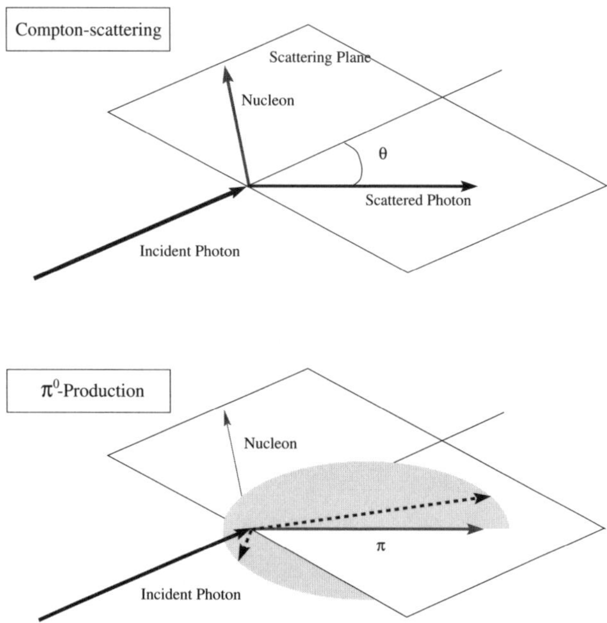

Fig. A.6. Kinematical view of Compton scattering and π^0 photoproduction. *Top*: the scattering reaction defines the scattering plane, in which the scattered photon is detected at the angle θ. The struck nucleon is emitted in the same plane. *Bottom*: assuming a π^0 meson is produced in the scattering plane, an immediate decay into two photons in the laboratory frame is allowed in accordance with the ellipsoid indicated

Using the equations above, we obtain the following for the emission angle of the recoiling nucleon:

$$\tan \vartheta_{\mathrm{N}} = \frac{-\omega' \sin \theta_\gamma}{\omega - \omega' \cos \theta_\gamma} \; . \tag{A.32}$$

It can be shown that in the laboratory system the nucleon recoil angle ϑ_{N} is less or equal to $90°$.

From the kinematical point of view, the scattering reaction is completely determined by two kinematical quantities. These can be for example, the incident photon energy and the scattering angle. Thus the identification of this reaction is rather simple. But above 145 MeV it is π^0 photoproduction which dominates the reaction cross section. Owing to the immediate 2γ decay, this also leads to photons in the final state. In the rest frame of the π^0, the 2γ decay is isotropic and the two photons share the mass of the meson, so that $\omega_1 = \omega_2 = m_\pi/2$, and they are emitted in opposite directions, i.e. $\Phi_{\gamma\gamma} = 180°$. In the laboratory frame, this is also exactly valid at threshold.

At higher incident photon energies, the π^0 energy also increases. Transforming the decay into the laboratory system leads to a totally different

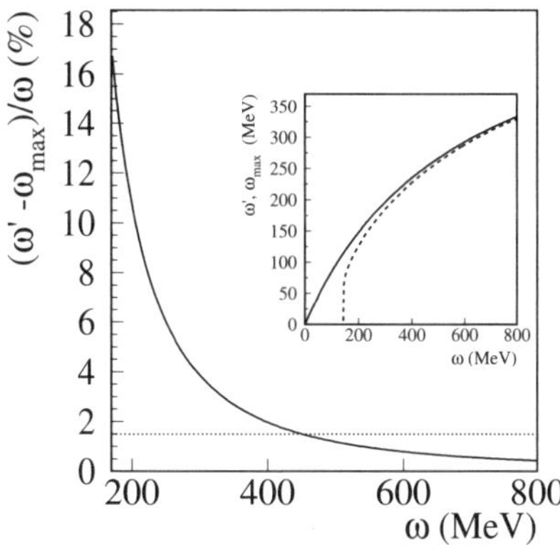

Fig. A.7. The relative difference between the energy of a scattered photon and the maximum energy of a π^0 decay photon as given by (A.34) for $\theta_\gamma = 130°$. The *dotted line* at 1.5% indicates the energy resolution of the CATS NaI(Tl) detector (Fig. A.13). The inset shows the absolute energy of a scattered photon (*solid line*) and the maximum energy of a π^0-decay photon (*dashed line*)

situation. The Lorentz boost transforms the photon momenta into the direction of the π^0 momentum and the decay is no longer isotropic. Thus the two photon momenta have to end on an ellipse, as indicated in Fig. A.6. The opening angle between the photons is now distributed between 180° and a minimum opening angle

$$\Phi_{\gamma\gamma}^{\min} = 2 \arcsin\left(\frac{\sqrt{E_\pi^2 - m_\pi^2}}{E_\pi}\right) , \tag{A.33}$$

where m_π is the π^0 mass and E_π its total energy. At this opening angle, both decay photons have the same energy: $\omega_1 = \omega_2 = E_\pi/2$. In the other extreme case, where $\Phi_{\gamma\gamma} = 180°$, a photon with the maximum energy $\omega_{\max}$ is emitted in the direction of the π^0 momentum and a low-energy photon with the minimum energy $\omega_{\min}$ is emitted in the opposite direction:

$$\omega_{\max} = \frac{1}{2}\left(E_\pi + \sqrt{E_\pi^2 - m_\pi^2}\right) , \tag{A.34}$$

$$\omega_{\min} = \frac{1}{2}\left(E_\pi - \sqrt{E_\pi^2 - m_\pi^2}\right) . \tag{A.35}$$

Assuming that the π^0 meson is emitted within the scattering plane (Fig. A.6), there is a good chance of detecting the high-energy decay photon. The energy of a scattered photon and a π^0 decay photon with the maximum energy becomes comparable above 400 MeV incident photon energy (Fig. A.7). There is no possibility of separating the two reactions via the emitted photon energy with the available energy resolutions of photon detectors. This may be overcome if both the photon and the recoiling proton are detected. For Compton scattering, the scattered photon and the recoiling proton have to be within the scattering plane (in plane), whereas for π^0 photoproduction the final particles can also be emitted outside the scattering plane (out of plane). Ideally, the out-of-plane distribution should be measured over a wide angular range above and below the scattering plane. Extrapolating into the scattering plane permits one to subtract the π^0 contribution from the measured in-plane distribution, and thus only photons from the Compton scattering process remain.

This technique of in-plane/out-of-plane subtraction has been used by the COPP experiment [17] in the Δ-energy range. It has been applied at higher energies by the LARA experiment [18].

A.6 Kinematical Description
of Quasi-Free Compton Scattering

The reaction $\gamma d \to \gamma np$ has to be considered in the case where the scattered photon and one nucleon, i.e. either the proton or the neutron, are detected in the final state. The remaining nucleon is then called the "spectator". Let $\boldsymbol{p}_\gamma$ and $\boldsymbol{p}'_\gamma$, $\boldsymbol{p}_{\mathrm{p}}$, $\boldsymbol{p}_{\mathrm{n}}$ be the three-momenta of the initial and final particles in the laboratory system (Fig. A.8).

The energy and momentum conservation conditions read as

$$\omega - \Delta = \omega' + E_{\mathrm{p}} + E_{\mathrm{n}} , \tag{A.36}$$

$$\boldsymbol{p}_\gamma = \boldsymbol{p}'_\gamma + \boldsymbol{p}_{\mathrm{p}} + \boldsymbol{p}_{\mathrm{n}} , \tag{A.37}$$

where $E_{\mathrm{p,n}}$ are the kinetic energies of the nucleons and $\Delta = 2.225$ MeV is the deuteron binding energy. Therefore, if the directions of the scattered photon $(\theta_\gamma, \Phi_\gamma)$ and of the recoiling nucleon $(\vartheta_{\mathrm{N}}, \Phi_{\mathrm{N}})$ are measured and in addition two of the energies ω, ω' and E_{N} are known, one can eliminate the momentum of the spectator nucleon.

In the case of quasi-free scattering, i.e. where the kinetic energy of the spectator nucleon is zero, all momenta of the particles lie in the same plane. In this case the kinematical relations of the reaction are comparable to the case of free scattering (Sect. A.5). For example, the energy of the scattered photon can be calculated from

$$\omega' = \frac{\omega(m_{\mathrm{N}} - \Delta - E_{\mathrm{N}} + |\boldsymbol{p}_{\mathrm{N}}| \cos \vartheta_{\mathrm{N}}) - (2m_{\mathrm{N}} - \Delta)(E_{\mathrm{N}} + \Delta/2)}{\omega(1 - \cos \theta_\gamma) + m_{\mathrm{N}} - \Delta - E_{\mathrm{N}} + |\boldsymbol{p}_{\mathrm{N}}| \cos \theta_{\gamma \mathrm{N}}} , \tag{A.38}$$

where N = p, n and $\theta_{\gamma \mathrm{N}}$ is the angle between the scattered photon and the recoiling nucleon. Thus, the difference between the expected photon energy and the measured value, i.e. the photon missing energy, can be calculated.

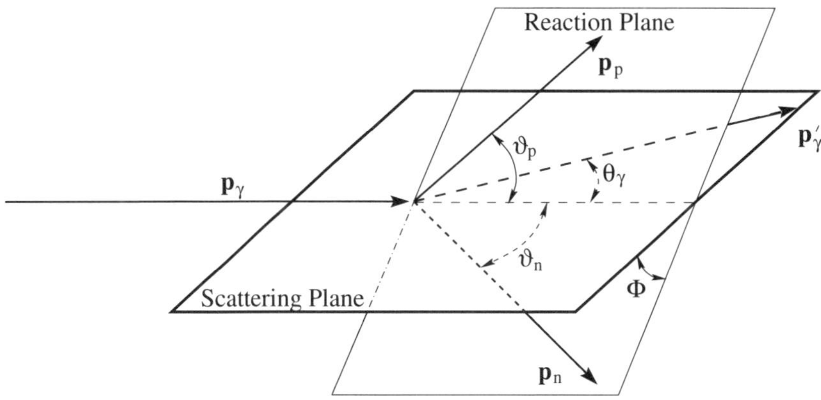

Fig. A.8. Outline of the kinematical variables of the reaction $\gamma d \rightarrow \gamma np$. The incident and scattered photons define the scattering plane. The nucleons are emitted within the reaction plane

The kinematical variables of the recoiling nucleon may also be calculated in the center of the quasi-free peak:

$$E_\mathrm{N}^\mathrm{qf} = \frac{\omega(\omega - \Delta)(1 - \cos\theta_\gamma) + \Delta^2/2}{m_\mathrm{N} - \Delta + \omega(1 - \cos\theta_\gamma)} \, , \tag{A.39}$$

$$\omega'^\mathrm{qf} = \omega - E_\mathrm{N}^\mathrm{qf} - \Delta \, , \tag{A.40}$$

$$\tan\vartheta_\mathrm{N}^\mathrm{qf} = \frac{-\omega'^\mathrm{qf}\sin\theta_\gamma}{\omega - \omega'^\mathrm{qf}\cos\theta_\gamma} \, . \tag{A.41}$$

For $\Delta = 0$, the above equations are the same as those describing the scattering from a free nucleon.

A.7 TAPS Detector

TAPS consisted of hexagonally shaped BaF_2 modules (5.9 cm diameter $\times$ 25 cm) [19], 64 of which were mounted in a block. Six blocks were placed around the target at distances of about 55 cm. Each module was equipped with an individual veto counter at the front to identify charged particles. An additional 120 BaF_2 modules were mounted in the forward direction symmetrically around the photon beam to form a forward wall (FW). An outline of the TAPS setup is given in Fig. A.9. The FW modules were of phoswich type, with a plastic veto counter mounted directly onto the crystal. The light output from both parts was collected by the same photomultiplier tube. This gives excellent identification of charged particles in addition to the pulse shape discrimination provided by BaF_2.

The main purpose of TAPS was detect the 2γ decay of neutral mesons, i.e. π^0 and η mesons. To reduce the readout frequency and enhance the content

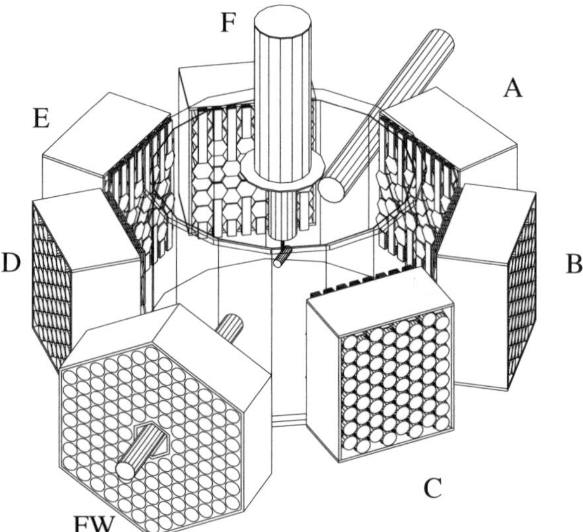

Fig. A.9. Setup of the TAPS detector system at the photon beam at MAMI. The photon beam enters the target chamber between the blocks A and F and leaves the setup through the forward wall (FW). The target at the center of the scattering chamber, the beam pipes for the photon beam and the cryogenic system on the top of the target chamber are shown. With permission from the TAPS collaboration

of the 2γ events, a standard trigger condition was defined in such a way that a minimum of two blocks, or one block and the FW, had to create the trigger signal. As an example of the detected 2γ spectrum, the invariant mass

$$M_{\mathrm{inv}} = \sqrt{2E_{\gamma 1}E_{\gamma 2}\left(1 - \cos\phi_{12}\right)} \tag{A.42}$$

of such events is shown in Fig. A.10. Here, $E_{\gamma 1}$ and $E_{\gamma 2}$ are the measured energies of the decay photons, and ϕ_{12} is the measured opening angle. The two photons were selected under the condition that the time difference between the two photons was within an interval of 0.7 ns. The agreement between the simulated and the measured response was optimized during several years of experience with the TAPS system [20]. The simulated response function for photons was optimized by a comparison between measured invariant-mass spectra of the kind shown in Fig. A.10 and simulated spectra.

The quasi-free experiment (Sect. 5.3.1) used only blocks A and F to detect the scattered photons and used the FW to detect the recoiling protons. The detection threshold of the FW detectors did not allow quasi-free scattering from the neutron to be included as well. Free Compton scattering at low photon energies (Sect. 4.1.1) was measured using the six blocks A to D only.

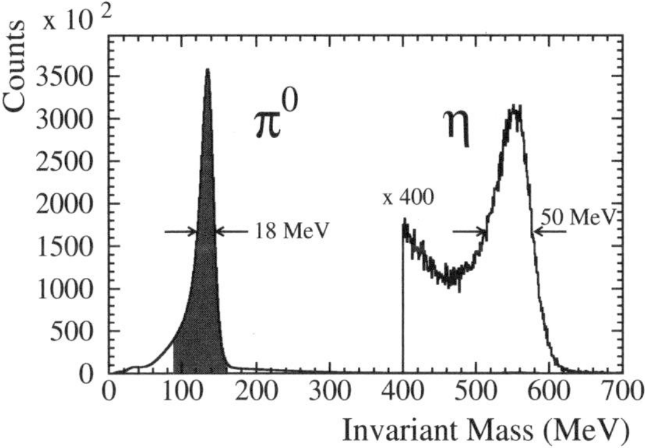

Fig. A.10. Example of the measured invariant-mass spectrum of 2γ events. Two clear, distinct peaks can be identified, corresponding to the π^0 meson at 135 MeV and the η meson at 550 MeV

The trigger condition for this experiment was set in such a way that each block was able to create a trigger signal.

A.8 CATS NaI(Tl) Detector

The **C**ompton **A**nd **T**wo photon **S**pectrometer (CATS) [21, 22] consists of the CATS NaI(Tl) detector, which detects the high-energy photons, and a 2π array of 61 BaF$_2$ detectors of the TAPS type (as described in Sect. A.7), which detects the low-energy photons from the $\pi^0 \rightarrow \gamma\gamma$ decay simultaneously. The large-volume (48 cm diameter $\times$ 64 cm) CATS NaI(Tl) detector (Fig. A.11) was used in a single-arm experiment to investigate Compton scattering from ^{12}C [23, 24] for the first time. Owing to the excellent photon energy resolution, differential cross sections for inelastic scattering into the first excited state of ^{12}C could be obtained.

The CATS NaI(Tl) detector[1] is of cylindrical shape and consists of a 27 cm diameter core crystal, for which two crystals of lengths 48 cm and 16 cm were cemented together. The core is surrounded by a segmented annulus of six optically separated crystals. The core crystal is viewed by seven (7.6 cm diameter) photomultiplier tubes (PMTs)[2], and each segment is viewed by four PMTs, two on each side. All crystal parts were carefully surface-compensated during manufacture to ensure uniform light collection. In order to suppress cosmic-ray-induced events, the NaI(Tl) detector is surrounded by six plastic scintillators: an annulus of five segments and a disk at the rear side. This anti-cosmic-ray shield has an efficiency of about 99%. Layers of ^{6}LiCO$_3$ 1 cm thick are inserted between the scintillators and the NaI(Tl)

---

[1] Manufactured by BICRON.
[2] Hamamatsu R1911.

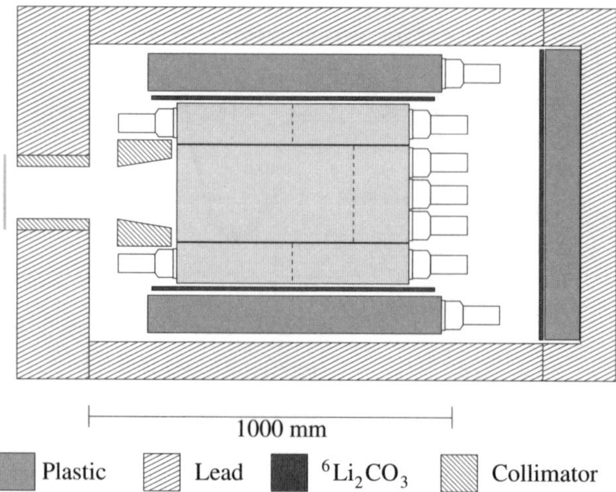

Fig. A.11. Outline of the CATS NaI(Tl) detector. With permission from Institut für Kernphysik, Universität Mainz

detector to absorb thermal neutrons. All detector parts are mounted in a lead housing. Owing to the collimation system, the scattered photons hit only the core crystal. Therefore the six segments detect the shower leakage out of the core.

The energy calibration of the CATS NaI(Tl) detector was performed in two consecutive steps: (i) the core was calibrated in the direct photon beam by making use of the energy calibration provided by the tagger (Sect. A.4), and (ii) the segments were calibrated using γ rays from an ^{241}Am–^{9}Be source positioned inside the collimator. The source delivers 4.43 MeV photons, and neutrons. The neutrons are thermalized and captured by hydrogen in the surrounding material, giving rise to additional photons of energy 2.23 MeV. The total information from the detector is the sum of the measured energies in all annular NaI(Tl) segments and in the core. The calibration measurements in the direct photon beam were also used to determine the response functions of the detector (Fig. A.12). For this purpose, the photon energy spectrum of the detector as a function of the missing energy $\Delta E_\gamma = x = \omega - E_{\mathrm{NaI}}$ was described by a Gaussian function and an exponential tail:

$$R(x) = N(g(x) + t(x)) \,,$$

$$g(x) = \exp\left(-\frac{(x - x_0)^2}{2\sigma^2}\right) \,,$$

$$t(x) = \begin{cases} \exp\left(-\frac{(x-x_0)}{\tau}\right)(1 - g(x)) & \text{for } (x - x_0) \geq 0.2\text{ MeV} \\ 0 & \text{for } (x - x_0) < 0.2\text{ MeV} \end{cases} \quad \text{.(A.43)}$$

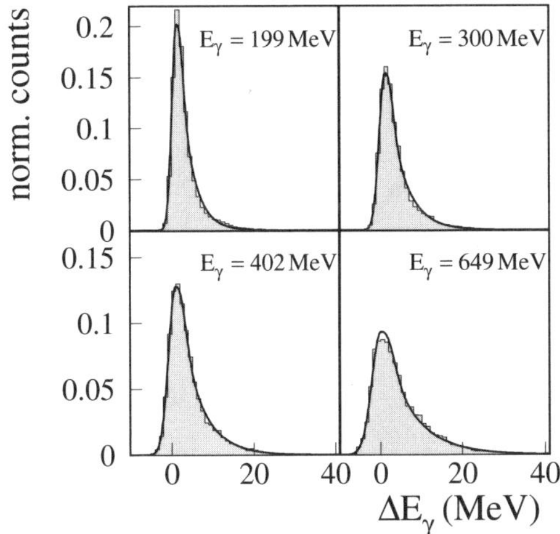

Fig. A.12. Example of the measured response functions of the CATS NaI(Tl) detector as a function of the missing energy $\Delta E_\gamma = \omega - E_{\mathrm{NaI}}$, which is the difference between the incident photon energy as obtained from the tagger and the photon energy measured with the CATS NaI(Tl) detector. The *shaded areas* are the measured spectra. The *solid lines* are the results of the fitting procedure using (A.43)

From these measurements, an energy-dependent set of parameters, i.e. $\sigma(x)$ and $\tau(x)$, could be obtained, which was then used within a Monte Carlo simulation in order to obtain the response functions for the scattered photons.

From the measured response functions of the kind shown in Fig. A.12, the intrinsic relative photon energy resolution, i.e. $\Delta E_{\mathrm{NaI}}/\omega$, can be extracted by deconvoluting the energy spread of the tagger $\Delta\omega$. This was done by assuming that the response of the tagger was rectangular, which corresponds to a variance $\sigma_\gamma^2 = \Delta\omega^2/12$. If we approximate the detector response with a Gaussian function, the variance of the response function is

$$\sigma \simeq \frac{\Delta E}{2.35}, \tag{A.44}$$

where ΔE is the full-width half-maximum of the measured response function. Assuming that the resulting variance σ^2 of the response function is the sum of the variances of the two contributions,

$$\sigma^2 = \sigma_\gamma^2 + \sigma_{\mathrm{NaI}}^2, \tag{A.45}$$

the intrinsic relative energy resolution may be determined from:

$$\left(\frac{\Delta E_{\mathrm{NaI}}}{\omega}\right)^2 = \left(\frac{\Delta E}{\omega}\right)^2 - \frac{1}{12}\left(\frac{2.35\Delta\omega}{\omega}\right)^2. \tag{A.46}$$

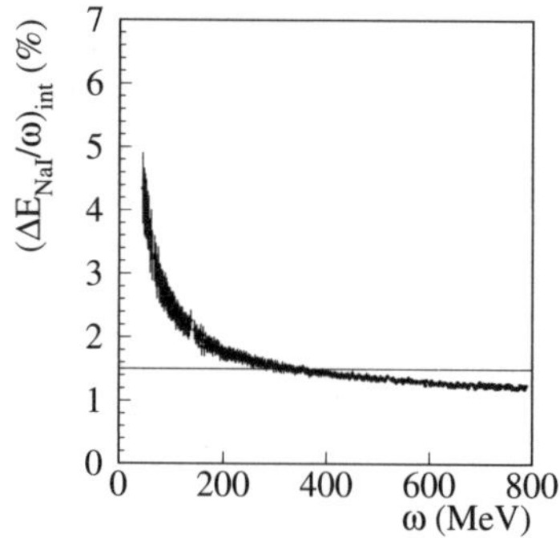

Fig. A.13. The measured intrinsic relative energy resolution of the CATS NaI(Tl) detector as a function of the photon energy

The result of such a measurement is shown in Fig. A.13. Above 250 MeV, the photon energy resolution of the CATS NaI(Tl) detector is better than 1.5%.

A.9 LARA Detector

The **LAR**ge **A**cceptance arrangement (LARA) experiment [18, 25, 26, 27, 28] (Fig. A.14) covered an angular range of 30° to 150°. This was achieved with 150 ($15 \times 15 \times 30$ cm^3) Pb glass detectors, which were set up in 30 stacks of five detectors each, on a semicircle with a radius of about 2 m around the target center. The detectors were stabilized with pulses of laser light. Therefore, any changes to the threshold calibrations could be monitored. Three stacks were covered by a plastic scintillator (veto detector, thickness 1 cm) at the front to identify charged particles. The very poor energy resolution of the Pb glass detectors, of about 30%, was compensated by measuring the trajectory and energy of the recoiling proton with the necessary resolution.

The proton trajectory was determined by two multiwire proportional chambers (MWPCs) at distances of 25 cm and 50 cm from the target center, and eight trigger detectors (thickness 0.5 cm) behind the wire chambers. The geometrical shape of the MWPCs was adapted to the requirements of the scattering kinematics. To identify protons at very small emission angles, i.e. as small as 6°, the MWPCs were constructed in such way that the photon beam itself had to pass through a wire-free area. Argon gas, enriched with a small fraction of isopropanol (C_3H_7OH) and isobutane (C_4H_{10}) in a ratio of 2:1, was chosen as the gas filling. In order to minimize the data readout, groups of trigger detectors in coincidence with the appropriate groups of Pb glass stacks, chosen in accordance with the scattering kinematics, triggered the readout. Thus, the number of protons from the scattering reaction was optimized relative to all other background reactions.

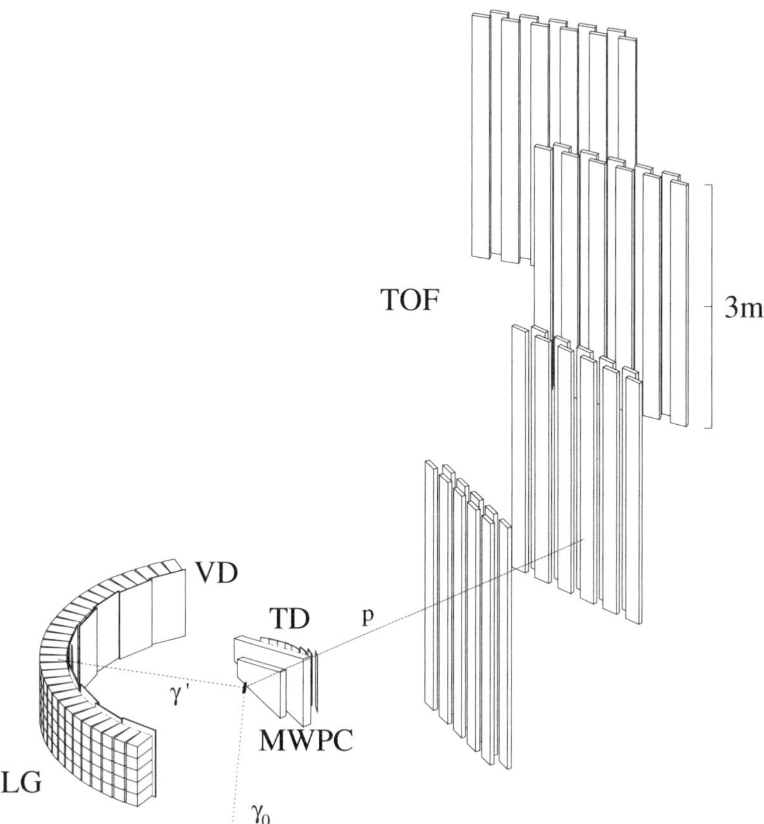

Fig. A.14. Outline of the LARA experiment. The incident photons hit the liquid-hydrogen target inside the scattering chamber (not shown). The scattered photons were detected by 150 Pb glass detectors (LG) mounted in 30 stacks of five detectors each. Three stacks were supplied with a veto detector (VD) in front to identify charged particles. The recoiling protons passed through two wire chambers (MWPC) and trigger detectors (TD) and reached the TOF detectors

The proton energy was determined by the time-of-flight (TOF) technique. The protons were detected with 43 plastic scintillators with dimensions $20 \times 5 \times 300$ cm^3 [29]; the scintillators were mounted on four frames in groups of 10 (one frame) or 11 (three frames). The TOF frames were set up at distances chosen in accordance with the various expected times of flight. Each TOF detector was equipped with a photomultiplier tube at each end. Thus, the time difference between the ends allowed one to recalculate the point where a proton had hit the detector. This allowed the experimenters to check the proton trajectory as determined by the MWPCs.

References

1. G. F. Chew et al., Phys. Rev. **106** (1957) 1345
2. A. Nagl, V. Devanathan, H. Überall, *Nuclear Pion Photoproduction*, STMP, Vol. 120, Springer, Berlin, Heidelberg, 1991
3. R. L. Walker, Phys. Rev. **182** (1969) 1729
4. R. A. Arndt et al., Phys. Rev. C **42** (1990) 1853
5. R. L. Workman, R. A. Arndt, Phys. Rev. D **45** (1992) 1789
6. R. E. Prange, Phys. Rev. **100** (1958) 240
7. A. I. L'vov, Sov. J. Nucl. Phys. **34** (1981) 597
8. A. I. L'vov, V. A. Petrun'kin, M. Schumacher, Phys. Rev. C **55** (1997) 359
9. A. C. Hearn, E. Leader, Phys. Rev. **126** (1962) 789
10. D. Babusci et al., Phys. Rev. C **58** (1998) 1013
11. W. A. Bardeen, W. K. Tung, Phys. Rev. **173** (1968) 1423
12. I. Anthony et al., Nucl. Instrum. Methods A **301** (1991) 230
13. S. J. Hall et al., Nucl. Instrum. Methods A **368** (1996) 698
14. H. Herminghaus, *Proc. of Linear Accelerator Conf.*, Albuquerque, NM, USA, 1990
15. T. Walcher, Prog. Part. Nucl. Phys. **24** (1990) 189
16. J. Ahrens, Institut für Kernphysik, Universität Mainz, private communication
17. C. Molinari et al., Phys. Lett. B **371** (1996) 181
18. S. Wolf, Dissertation, Universität Göttingen (1998), Cuvillier, Göttingen, 1998
19. R. Novotny, IEEE Trans. Nucl. Sci. **38** (1991) 379
20. A. R. Gabler et al., Nucl. Instrum. Methods A **346** (1994) 168
21. J. Peise et al., Phys. Lett. B **384** (1996) 37
22. A. Hünger et al., Nucl. Phys. A **620** (1997) 385
23. F. Wissmann, Dissertation, Universität Mainz (1993)
24. F. Wissmann et al., Phys. Lett. B **335** (1994) 119
25. G. Galler, Dissertation, Universität Göttingen (1998), Cuvillier, Göttingen, 1998
26. G. Galler et al., Phys. Lett. B **503** (2001) 245
27. S. Wolf et al., Eur. Phys. J. A **12** (2001) 231
28. V. Lisin, Institute of Nuclear Research, Moscow, private communication
29. P. Grabmayr et al., Nucl. Instrum. Methods A **402** (1998) 85

Index

Classified Index

Springer Tracts in Modern Physics, Volumes 36–200

This cumulative index is based upon the Physics and Astronomy classification Scheme (PACS) developed by the American Institute of Physics

General

02.30 Function Theory, Analysis

02.70 Computational Techniques

03.50 Classical Field Theories

03.65 Quantum Mechanics

03.67 Quantum Information

04 Relativity and Gravitation

04.62 Quantum Field Theory in Curved Spacetime

05 Statistical Physics

05.10 Computational Methods in Statistical Physics and Nonlinear Dynamics

Heinloth, K.: Experiments on Electroproduction in High Energy Physics (Vol. 65)

Höhler, G.: Special Models and Predictions for Pion Photoproduction (Low Energies) (Vol. 39)

von Holtey, G.: Pion Photoproduction on Nucleons in the First Resonance Region (Vol. 59)

Katz, U. F.: Deep Inelastic Positron–Proton Scattering in the High-Momentum-Transfer Regime of HERA (Vol. 168)

Kolanoski, H.: Two-Photon Physics at e^+e^- Storage Rings (Vol. 105)

Landshoff, P.V.: Duality in Deep Inelastic Electroproduction (Vol. 62)

Llewellyn Smith, C.H.: Parton Models of Inelastic Lepton Scattering (Vol. 62)

Lücke, D., Söding, P.: Multipole Pion Photoproduction in the s-Channel Resonance Region (Vol. 59)

Osborne, L.S.: Photoproduction of Mesons in the GeV Range (Vol. 39)

Pfeil, W., Schwela, D.: Coupling Parameters of Pseudoscalar Meson Photoproduction on Nucleons (Vol. 55)

Renard, F.M.: ϱ-ω Mixing (Vol. 63)

Rittenberg, V.: Scaling in Deep Inelastic Scattering with Fixed Final States (Vol. 62)

Rollnik, H., Stichel, P.: Compton Scattering (Vol. 79)

Rubinstein, H.R.: Duality for Real and Virtual Photons (Vol. 62)

Rühl, W.: Application of Harmonic Analysis to Inelastic Electron–Proton Scattering (Vol. 57)

Schildknecht, D.: Vector Meson Dominance, Photo- and Electroproduction from Nucleons (Vol. 63)

Schilling, K.: Some Aspects of Vector Meson Photoproduction on Protons (Vol. 63)

Schwela, D.: Pion Photoproduction in the Region of the Δ (1230) Resonance (Vol. 59)

Wissmann, F.: Compton Scattering (Vol. 200)

Wolf, G.: Photoproduction of Vector Mesons (Vol. 57)

13.66 Lepton–Lepton Interactions

Altarelli, G., Winter, K.: Neutrino Mass (Vol. 190)

Kilian, W.: Electroweak Symmetry Breaking (Vol. 198)

Kolanoski, H.: Two-Photon Physics at e^+e^- Storage Rings (Vol. 105)

Kramer, G.: Theory of Jets in Electron–Positron Annihilation (Vol. 102)

Kuznetsov, A., Mikheev, N.: Electroweak Processes in External Electromagnetic Fields (Vol. 197)

Stahl, A.: Physics with Tau Leptons (Vol. 160)

13.75 Hadron-Induced Reactions

Atkinson, D.: Some Consequences of Unitarity and Crossing. Existence and Asymptotic Theorems (Vol. 57)

Basdevant, J. L.: $\pi\pi$ Theories (Vol. 61)

DeSwart, J.J., Nagels, M.M., Rijken, T.A., Verhoeven, P.A.: Hyperon–Nucleon Interaction (Vol. 60)

Ebel, G., Julius, D., Kramer, G., Martin, B. R., Müllensiefen, A., Oades, G., Pilkuhn, H., Pišút, J., Roos, M., Schierholz, G., Schmidt, W., Steiner, F., DeSwart, J.J.: Compilation of Coupling Constants and Low-Energy Parameters (Vol. 55)

Gustafson, G., Hamilton, J.: The Dynamics of Some π–N Resonances (Vol. 57)

Hamilton, J.: New Methods in the Analysis of πN Scattering (Vol. 57)

Kramer, G.: Nucleon–Nucleon Interactions below 1 GeV/c (Vol. 55)

Lichtenberg, D.B.: Meson and Baryon Spectroscopy (Vol. 36)

Martin, A.D.: The Λ KN Coupling and Extrapolation below the KN Threshold (Vol. 55)

Martin, B.R.: Kaon–Nucleon Interactions below 1 GeV/c (Vol. 55)

Morgan, D., Pišút, J.: Low Energy Pion–Pion Scattering (Vol. 55)

Oades, G.C.: Coulomb Corrections in the Analysis of πN Experimental Scattering Data (Vol. 55)

Pišút, J.: Analytic Extrapolations and the Determination of Pion–Pion Shifts (Vol. 55)

Wanders, G.: Analyticity, Unitarity and Crossing-Symmetry Constraints for Pion–Pion Partial Wave Amplitudes (Vol. 57)

Zinn-Justin, J.: Course on Padé Approximants (Vol. 57)

of State. In Honor of Friedrich Hund's 100th Birthday (Vol. 133)

23.00 Weak Interactions

Gasiorowicz, S.: A Survey of the Weak Interaction (Vol. 52)
Primakoff, H.: Weak Interactions in Nuclear Physics (Vol. 53)

23.00 Radioactive Decay and In-Beam Spectroscopy

Andreev, B.M., Magomedbekov, E.P., Sicking, G.H.: Interaction of Hydrogen Isotopes with Transition-Metals and Intermetallic Compounds (Vol. 132)

23.40 Beta Decay; Double Beta Decay; Electron and Muon Capture

Faessler, A., Kosmas, T.S., Leontaris, G.K. (Eds.): Symmetries in Intermediate and High Energy Physics (Vol. 163)

25.20 Photonuclear Reactions

Nagl, A., Devanathan, V., Überall, H.: Nuclear Pion Photoproduction (Vol. 120)
Wissmann, F.: Compton Scattering (Vol. 200)

25.30 Lepton-Induced Reactions and Scattering

Theißen, H.: Spectroscopy of Light Nuclei by Low Energy (70 MeV) Inelastic Electron Scattering (Vol. 65)
Überall, H.: Electron Scattering, Photoexcitation and Nuclear Models (Vol. 49)

25.40 Nucleon-Induced Reactions

Lechner, R.E., Richter, D. Riekel C.: Neutron Scattering and Muon Spin Rotation (Vol. 101)

25.80 Meson- and Hyperon-Induced Reactions

Nagl, A., Devanathan, V., Überall, H.: Nuclear Pion Photoproduction (Vol. 120)

28.20 Neutron Physics

Koestner, L.: Neutron Scattering Lengths and Fundamental Neutron Interactions (Vol. 80)
Springer, T.: Quasi-Elastic Scattering of Neutrons for the Investigation of Diffusive Motions in Solids and Liquids (Vol. 64)
Steyerl, A.: Very Low Energy Neutrons (Vol. 80)

Andreev, B.M., Magomedbekov, E.P., Sicking, G.H.: Interaction of Hydrogen Isotopes with Transition-Metals and Intermetallic Compounds (Vol. 132)

28.60 Isotope Separation and Enrichment

Ehrfeld, W.: Elements of Flow and Diffusion Processes in Separation Nozzles (Vol. 97)

29 Experimental Methods

Panofsky, W.K.H.: Experimental Techniques (Vol. 39)
Strauch, K.: The Use of Bubble Chambers and Spark Chambers at Electron Accelerators (Vol. 39)

29.20 Cyclic Accelerators and Storage Rings

Kolanoski, H.: Two-Photon Physics at e^+e^- Storage Rings (Vol. 105)

29.25 Particle Sources and Targets

Kleinknecht K., Dee, T.D. (Eds.): Particles and Detectors. Festschrift for Jack Steinberger (Vol. 108)

29.27 Beams in Particle Accelerators

Kirschner J.: Polarized Electrons at Surfaces (Vol. 106)

29.30 Spectrometers and Spectroscopic Techniques

Kleinknecht, K.: Uncovering CP Violation (Vol. 195)

29.40 Radiation Detectors

Katz, U.F.: Deep Inelastic Positron–Proton Scattering in the High-Momentum-Transfer Regime of HERA (Vol. 168)
Kleinknecht, K.: Uncovering CP Violation (Vol. 195)
Sitar, B. Merson, G.I., Chechin, V.A., Budagov, Yu.A.: Ionization Measurements in High Energy Physics (Vol. 124)

Atomic and Molecular Physics

31.00 Electronic Structure of Atoms and Molecules, Theory

Donner, W., Süßmann, G.: Paramagnetische Felder am Kernort (Vol. 37)

Electromagnetism, Optics, Acoustics, Heat Transfer, Classical Mechanics and Fluid Mechanics

81.40 Treatment of Materials and Its Effects on Microstructure and Properties

Hein, M.: High-Temperature-Superconductor Thin Films at Microwave Frequencies (Vol. 155)

82.30 Specific Chemical Reactions; Reactions Mechanism

Khabibullaev, P.K., Skorodumov, B.G.: Determination of Hydrogen in Materials Nuclear Physics Methods (Vol. 117)

82.55 Radiochemistry

Andreev, B.M., Magomedbekov, E.P., Sicking, G.H.: Interaction of Hydrogen Isotopes with Transition-Metals and Intermetallic Compounds (Vol. 132)

82.60 Chemical Thermodynamics

Ledentsov, N.N.: Growth Processes and Surface Phase Equilibria in Molecular Beam Epitaxy (Vol. 156)

82.65 Surface and Interface Chemistry: Heterogeneous Catalysis at Surfaces

Pockrand, I.: Surface Enhanced Raman Vibrational Studies at Solid/Gas Interfaces (Vol. 104)
Raether, H.: Surface Plasmons on Smooth and Rough Surfaces and on Gratings (Vol. 111)

82.70 Disperse Systems; Complex Fluids

Bibette, J., Leal-Calderon, F., Schmitt, V., Poulin, P.: Emulsion Science (Vol. 181)

82.80 Chemical Analysis and Related Physical Methods of Analysis

Fromme, B.: d-d Excitations in Transition-Metal Oxides (Vol. 170)

83.70 Material Form

Ristow, G.H.: Pattern Formation in Granular Materials (Vol. 164)

84.40 Radiowave and Microwave Technology

POAN Research Group: New Aspects of Electromagnetic and Acoustic Wave Diffusion (Vol. 144)

85.25 Superconducting Devices

Hein, M.: High-Temperature-Superconductor Thin Films at Microwave Frequencies (Vol. 155)

85.60 Vacuum Tubes

Ossicini, S., Pavesi, L., Priolo, F.: Light Emitting Silicon for Microphotonics (Vol. 194)

85.70 Magnetic Devices

Lehner, G.: Über die Grenzen der Erzeugung sehr hoher Magnetfelder (Vol. 47)

87.66 Radiation Measurements

Sitar, B. Merson, G.I., Chechin, V.A., Budagov, Yu.A.: Ionization Measurements in High Energy Physics (Vol. 124)

89.65 Urban Planning and Construction

Schulz, M.: Statistical Physics and Economics (Vol. 184)

89.75 Complex Systems

Schulz, M.: Statistical Physics and Economics (Vol. 184)
Staliūnas, K., Sánchez-Morcillo, V.: Transverse Patterns in Nonlinear Optical Resonators (Vol. 183)

Geophysics, Astronomy, and Astrophysics

91.30 Seismology

POAN Research Group: New Aspects of Electromagnetic and Acoustic Wave Diffusion (Vol. 144)

95.00 Theoretical Astrophysics

Kundt, W.: Recent Progress in Cosmology (Isotropy of 3 deg Background Radiation and Occurrence of Space-Time Singularities) (Vol. 47)
Kundt, W.: Survey of Cosmology (Vol. 58)
Stewart, J., Walker, M.: Black Holes: The Outside Story (Vol. 69)

95.30 Fundamental Aspects of Astrophysics

Kuznetsov, A., Mikheev, N.: Electroweak Processes in External Electromagnetic Fields (Vol. 197)

**95.55 Astronomical and
Space-Research Instrumentation**

Kleinknecht, K., Dee, T. D.(Eds.): Particles and Detectors Festschrift for Jack Steinberger (Vol. 108)

96.40 Cosmic Rays

Faessler, A., Kosmas, T. S., Leontaris, G. K. (Eds.) : Symmetries in Intermediate and High Energy Physics (Vol. 163)

96.60 Solar Physics

Riffert, H., Müther, H., Herold, H., Ruder, H.: Matter at High Densities in Astrophysics Compact Stars and the Equation of State. In Honor of Friedrich Hund's 100th Birthday (Vol. 133)

97.00 Stars

Börner, G.: On the Properties of Matter in Neutron Stars (Vol. 69)

97.60 Late Stages of Stellar Evolution

Kuznetsov, A., Mikheev, N.: Electroweak Processes in External Electromagnetic Fields (Vol. 197)

Riffert, H., Müther, H., Herold, H., Ruder, H.: Matter at High Densities in Astrophysics Compact Stars and the Equation of State. In Honor of Friedrich Hund's 100th Birthday (Vol. 133)

Springer Tracts in Modern Physics

Printing: Mercedes-Druck, Berlin
Binding: Stein+Lehmann, Berlin